Analysis of New Diesel Engine and Component Design

SP-1091

GLOBAL MOBILITY DATABASE

All SAE papers, standards, and selected books are abstracted and indexed in the Global Mobility Database.

Published by:
Society of Automotive Engineers, Inc.
400 Commonwealth Drive
Warrendale, PA 15096-0001
USA
Phone: (412) 776-4841
Fax: (412) 776-5760
February 1995

Permission to photocopy for internal or personal use, or the internal or personal use of specific clients, is granted by SAE for libraries and other users registered with the Copyright Clearance Center (CCC), provided that the base fee of $6.00 per article is paid directly to CCC, 222 Rosewood Drive, Danvers, MA 01923. Special requests should be addressed to the SAE Publications Group. 1-56091-641-9/95 $6.00.

No part of this publication may be reproduced in any form, in an electronic retrieval system or otherwise, without the prior written permission of the publisher.

ISBN 1-56091-641-9
SAE/SP-95/1091
Library of Congress Catalog Card Number: 94-74744
Copyright 1995 Society of Automotive Engineers, Inc.

Positions and opinions advanced in this paper are those of the author(s) and not necessarily those of SAE. The author is solely responsible for the content of the paper. A process is available by which discussions will be printed with the paper if it is published in SAE Transactions. For permission to publish this paper in full or in part, contact the SAE Publications Group.

Persons wishing to submit papers to be considered for presentation or publication through SAE should send the manuscript or a 300 word abstract of a proposed manuscript to: Secretary, Engineering Meetings Board, SAE.

Printed in USA

PREFACE

SAE special publication <u>Analysis of New Diesel Engine and Component Design</u> (SP-1091) contains research papers from authors worldwide. One new diesel engine is presented. Other papers investigate analysis techniques on components ranging from cylinder heads to vibration dampers. A new approach to piston cooling is explored as well as ring/liner tribology optimization with non-desirable fuel usage. Finally a new approach to reduce concept-to-customer time is offered.

John Whitacre
Zollner Pistons

Session Organizer

TABLE OF CONTENTS

950518   **A Root Cause Investigation of Cylinder Head Cracking in Large Diesel Engine Standby Power Generators** 1
     Keith D. Vertin
       Ricardo North America, Inc.
     Christian L. Haller and Thomas S. Lubnow, Jr.
       MPR Associates, Inc.

950520   **Predictive Analysis of Lube Oil Consumption for a Diesel Engine** 13
     Walter Zottin, Marcos Clemente, and José Manoel Martins Leite
       Metal Leve S.A. Ind. e Com.

950521   **Development of a New Engine Piston Incorporating Heat Pipe Cooling Technology** 19
     Qian Wang, Yiding Cao, and Alejandro Souto
       Florida International Univ.

950522   **Development of New Torsional Vibration Rubber Damper of Compression Type** 27
     Tomoaki Kodama, Yasuhiro Honda, and Katsuhiko Wakabayashi
       Kokushikan Univ.
     Shoichi Iwamoto
       Saitama Univ.

950524   **Novel Approach to Reduce the Time from Concept-to-Finished Piston** 45
     Jeffrey L. Castleman and Darren R. Bailey
       Zollner Corp.

950525   **Isuzu New V8 - V12 PE-Series Diesel Engines** 57
     Jiro Saito, Masaru Odajima, and Tetsuya Harada
       Isuzu Motors, Ltd.

950526   **Piston Cooling with Shaking-Up Heat Pipes (SUHP) and Thermal Analysis of the Cooling System** 73
     Yiding Cao and Qian Wang
       Florida International Univ.

950527   **The Effect of the Addition of Hard Particles on the Wear of Liner and Ring Materials Running with High Sulfur Fuel** 79
     Jan Vatavuk and Valmir Demarchi
       COFAP - Cia. Fabricadora de Peças

950518

A Root Cause Investigation of Cylinder Head Cracking in Large Diesel Engine Standby Power Generators

Keith D. Vertin
Ricardo North America, Inc.

Christian L. Haller and Thomas S. Lubnow, Jr.
MPR Associates, Inc.

ABSTRACT

Cylinder head cracking has been an engine development problem since the first high performance diesels were designed and manufactured in the early 20th century. Valve bridge cracking is a common failure mode that is very dependent on engine application and operating conditions. Cracking failures cause increased engine maintenance and downtime, costly part replacement and in rare cases catastrophic engine failure. Cylinder head cracking continues to be problematic for modern diesel engines as peak firing pressures increase to meet exhaust emissions legislation and BMEPs increase for improved power density.

The root cause of cylinder head cracking is often difficult to diagnose due to large number of design, manufacturing and engine operational variables involved. This paper summarizes the methods, results and conclusions of a study to determine the root cause of cracking in cylinder heads of large diesel engines used for standby power generation in nuclear plants. The root cause study included investigative, analytical and engine testing activities.

The study revealed that head cracking was primarily affected by the seat insert counterbore design, fire deck thickness, engine load, and casting anomalies adjacent to the cracked locations. Recommendations to extend the service life of the cylinder head are discussed.

INTRODUCTION

Cooper-Bessemer (C-B) Model KSV engines are installed at several nuclear power plants to serve as standby diesel generators. The KSV engine is a four stroke direct injection diesel with a 343x419mm bore and stroke. These engines are required to be available in the event of a loss of off-site power. To demonstrate this capability, the engines are periodically started and quickly taken to full load. The engines are rapidly started with compressed air within 10 seconds, and in some cases reach full load within 90 seconds.

Owners of KSV engines in nuclear standby service have found several cylinder heads to be cracked on the flame face surface. The cracks are normally observed to initiate between the pressed-in valve seat inserts. Cracks develop predominantly at the narrowest part of the exhaust valve bridge, between the two exhaust valve seat inserts as shown in Figure 1. In addition, some heads have been found to contain numerous small radial cracks around the counterbores for the exhaust seat inserts. Both cracking modes are shown in Figure 2. Most of the observed cracks are shallow surface cracks identified when the heads are removed for inspection. Shallow cracks of this type do not preclude satisfactory operation of the engine. In two cases, cracks have progressed through the fire deck. Through-wall cracks are considered unacceptable for use in service, as there would be increased risk of coolant leaking into the combustion chamber. Jacket water leakage could fill the combustion chamber during engine down time to the point that hydraulic locking of the cylinder and engine failure occurs during the next start attempt. Although only two cylinder heads with through-wall cracks have been reported in service, standard practice is to replace cylinder heads having surface cracks to maintain a low risk of catastrophic failure. Some cylinder heads have been refurbished after grinding out minor cracks.

The Cooper-Bessemer Owner's Group (CBOG) sponsored an investigation to determine the cause of cracking in KSV cylinder heads in order to identify appropriate corrective actions. CBOG consists of engine owners that use KSV engines in nuclear standby service. Other participants in the investigation included Cooper-Bessemer, MPR Associates, Ricardo North America, Commonwealth Edison, the Sumner Municipal Light Plant, and Preventive Maintenance Services.

---

* Numbers in brackets designate references at end of paper.

APPROACH

The intent of the root cause study was to understand the cylinder head cracking mechanism, review previous corrective actions, and identify further possible corrective actions to extend the service life of the cylinder head. The investigation was a collaborative effort including the following activities:

- Published literature was reviewed to investigate previous experiences with cylinder head cracking and the resolution of cracking failures.

- Cooper-Bessemer Reciprocating Products Division was consulted to review changes to the engine configuration, materials, and manufacturing processes of KSV cylinder heads over the thirty year history of the engine design.

- Cooper-Bessemer and other KSV engine operators were surveyed to determine the number of cracked heads, length of service before crack detection, severity of the observed cracks, and types of repairs to refurbish cylinder heads for use.

- KSV engine operation was reviewed at various installations to determine if there were any unusual operating conditions that may have led to increased cracking failure rates.

- KSV cylinder heads were inspected to reveal possible geometric factors which may affect cracking. The inspection consisted of visual inspections for cracks, video probe inspections of the internal coolant passages to identify casting defects or other flow restrictions, and ultrasonic measurements of the fire deck thickness at the valve bridge.

- A detailed metallurgical examination of a cracked head was performed. The purpose of this examination was to attempt to determine if material defects, material quality or geometric factors contributed to cracking.

- A finite element model of the cylinder head was developed to determine if operating temperatures and stresses were sufficient to cause cracking for nuclear standby applications. Sensitivity studies were subsequently performed to examine the effect of engine load, fire deck thickness variation, and design changes on cylinder head durability.

- Engine tests were performed to evaluate the sensitivity of fire deck temperatures to engine load and coolant flow rate, and to validate the finite element model.

Figure 1. A Crack that Propagated Across the Entire Width of the Exhaust Valve Bridge was Removed for Metallographic Examination.

RESULTS

HISTORICAL REVIEW - The valve bridge cracking failure mode evident on diesel cylinder heads is generally attributed to a low cycle thermal fatigue failure mechanism. The flame face of the cylinder head is exposed to hot combustion gasses in the cylinder. For a four valve head, peak temperatures usually occur

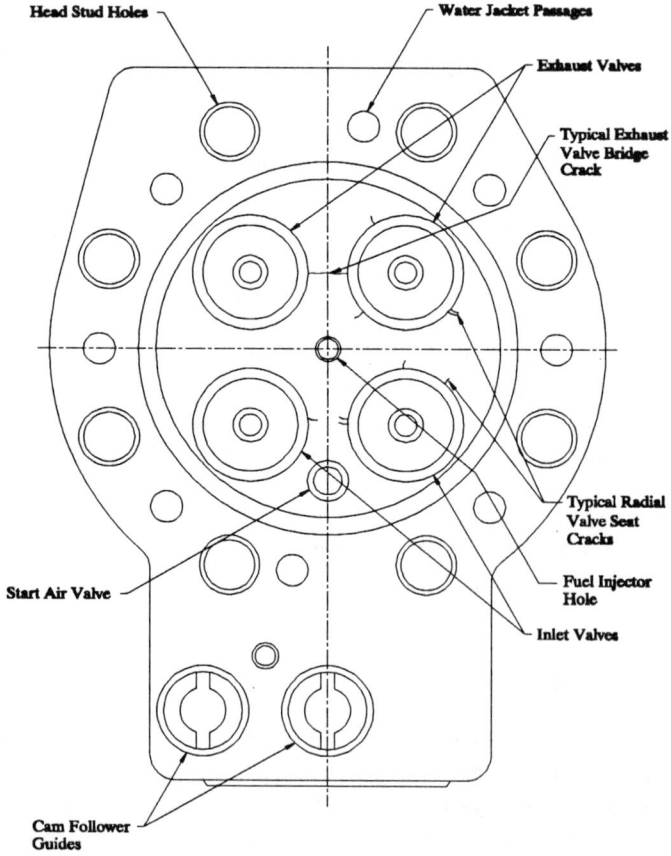

Figure 2. Typical Crack Locations Revealed by Inspections and Engine Owner Surveys

at the valve bridge between exhaust valves. The thermal expansion of the valve bridge is constrained by the lesser expansion of the bottom deck, which is significantly cooler outside the cylinder walls. Compressive stress develops at the valve bridge during constant load engine operation because the valve bridge is significantly hotter than its surroundings and full thermal expansion is prevented.

If the valve bridge becomes too hot during engine operation, compressive stresses become sufficiently large to cause permanent plastic deformation locally at the valve bridge, and the onset of the failure mechanism. Thermal fatigue damage results when the thermal load is reduced during engine load reduction or shut down. The bottom deck recovers elastically upon thermal load removal, but tensile stress develops in the valve bridge which has undergone plastic deformation. The valve bridge cycles between compressive and tensile states of stress during engine loading and unloading (or starts and stops) respectively, causing low cycle thermal fatigue damage and eventual failure. Additional fatigue damage may be caused by accumulated cylinder firing events or creep over long periods of continuous engine operation, resulting in shorter service life. A further review of cylinder head failure modes and mechanisms discussed in published literature was documented [1].

Cylinder head durability is clearly dependent on operating temperatures. Fatigue damage will result when operating temperatures produce thermal stresses that exceed the yield strength of the material. For gray iron castings (the same class of material used for KSV heads) a limiting operating temperature of 380°C was reported based on service experience [2]. The occurrence of cracking is reported to be almost assured for conventional head designs and materials if temperatures are greater than 400°C. Within this temperature range, variations in head material properties and design considerations (bridge width and thickness, valve seat insert press-fit, etc.) can influence the likelihood of cracking.

Published literature revealed several approaches to reduce cylinder head operating temperatures to reduce cracking. A small sampling follows.

- Directed coolant flow over the valve bridges is a proven way to reduce operating temperatures and cracking failures [3,4].

- Drilled passages have been recommended over the valve bridge as an alternative to "cast in place" passages [2]. This approach provides coolant flow close to the flame face surface and improved control of deck thickness variation. Drilled passages eliminate core clean-out and casting defects which could otherwise obstruct coolant flow. Heat transfer is improved across the drilled surface relative to an as-cast surface.

- Reduced fire deck thickness reduces thermal resistance and improves conductive heat transfer through the valve bridge [5]. A tradeoff exists between operating temperatures and stresses, since the section thickness is reduced.

- Application of thermal barrier coatings to insulate the cast iron fire deck has been used successfully in smaller engines to reduce flame face metal temperatures [4].

Other solutions to head cracking problems reportedly involve reduction of stresses resulting from thermal gradients. Those solutions principally involve reduction in the degree of constraint of the fire deck [6]. Design modifications proposed to achieve this objective include:

- Increasing the fire deck stiffness relative to that of the surrounding material,

- Machining scallops or slots into the fire deck to relieve stress,

- Installing a separate heat resistant fire deck which is not required to resist full engine firing pressure,

- Elimination of stress risers.

Proper engine operating procedures and maintenance practices are also essential to maintain low cylinder head failure rates. Any one-time overload event such as cylinder overpressurization, engine overheating, or temporary loss of engine cooling could produce sufficient fatigue damage to cause head cracking.

SURVEY AND INSPECTION - All KSV engines in nuclear service are operated at full load conditions with rapid starts every month and with rapid loading every six, twelve, or eighteen months as required by the plant technical specifications. KSV engines in nuclear service are required to operate at 110% of rated power

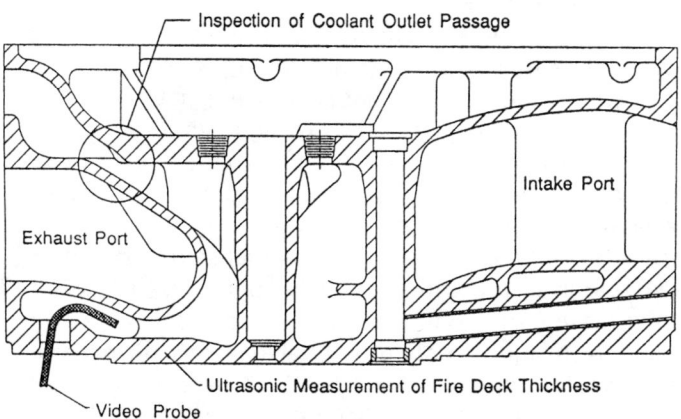

Figure 3. Features Examined During Cylinder Head Inspection

every eighteen months. The 110% load ratings for KSV engines in nuclear service range from 13.7 bar to 15.9 bar BMEP.

The survey revealed that engine ratings and peak firing pressures vary widely in service. Evidence of excessive peak cylinder pressures have been reported and have lead to valve bridge cracking. In one instance, fuel injection timing was advanced several degrees for 24 hours of operation due to slippage of the camshaft drive chain on the camshaft sprockets. Peak cylinder pressures were reported to be greater than twice the normal values. Eighteen of twenty cylinder heads were reportedly cracked following this event [7], indicating that mechanical loading may be part of the failure mechanism.

The rate of engine loading and unloading may affect the cylinder head cracking by increasing the magnitude of the thermal gradients in the casting. More rapid cycling results in increased likelihood for crack initiation and growth. Prior to 1992, many operators of KSV engines in the nuclear power industry did not follow a precise loading and unloading schedule during monthly surveillance testing. As a result, many engines were loaded and unloaded relatively quickly. The Cooper-Bessemer Technical Manual was revised in 1992 to incorporate a more gradual loading and unloading schedule. However, rapid loading (to 100% in 60 seconds) is still an operating procedure requirement at six to eighteen month intervals.

Inspection of KSV diesel engine cylinder heads revealed two types of cracking. The primary type of cracks were broad, long cracks across the entire width of the valve bridge as shown in Figure 1. All but a few of the cracks were located between the exhaust valve seat inserts. The secondary type of cracking was multiple, fine, short radial cracks distributed around the seat insert counterbores. This type of cracking is characteristic of thermal shock and is a common predecessor to valve bridge cracking. It is speculated that the primary cracking mode results at least in some cases from propagation of the short radial cracks across the valve bridge. Two cylinder heads had through-wall cracking which may have been caused by engine overheating.

For the heads inspected, the measured exhaust valve bridge thickness ranged from -10% to +40% of the design value with a majority of heads having a deck thickness greater than the design value. Metal penetration defects were observed in the majority of the cracked heads. This defect was due to molten metal seeping into the parting line in the sand casting core and solidifying. The defect is therefore difficult to remove during the core clean-out operation. In all but a few cases, the defects were minor and were not expected to significantly affect valve bridge stiffness or heat transfer capability. All of the large casting metal penetration defects were located above the exhaust valve bridge region. A significant variation in the coolant outlet passage above the exhaust ports was also noted (Figure 3). The cross-sectional flow area was found to be restrictive relative to the print dimensions in one case. This variation, along with the metal penetration casting defects, may have caused differences in coolant flow distribution among the cylinder heads.

Many of the cylinder heads from the inspection were used in applications other than nuclear standby power generation. Nevertheless, the casting quality was believed to be representative of most cylinder heads currently in service. The casting processes have been improved in recent years to improve dimensional control of the fire deck thickness, and to identify and eliminate other casting defects.

METALLURGICAL EVALUATION - A metallurgical evaluation was performed for a KSV cylinder head which developed through-deck valve bridge cracking in nuclear standby service. The evaluation consisted of sectioning a cracked head, internal and external visual examination of the head, microscopic examination of the microstructure and of the crack features, chemical analysis, and hardness and tensile strength testing [1].

Visual examination following head sectioning revealed a large metal penetration casting defect on the cooling water side of the fire deck adjacent to the crack. In addition, casting core shifts and significant discrepancies between the actual casting and drawing dimensions were observed. Chemical analyses and mechanical testing revealed that the alloy composition and material strength were in accordance with the material specification. The metallographic examination revealed that the casting microstructure was normal and of high quality for this type of casting with no microstructural defects found in the area of the crack. The location, orientation and general features of the crack were found to be suggestive of a fatigue cracking mechanism (Figure 4). However, oxidation of the crack faces precluded identification of features which could conclusively confirm the fatigue mechanism.

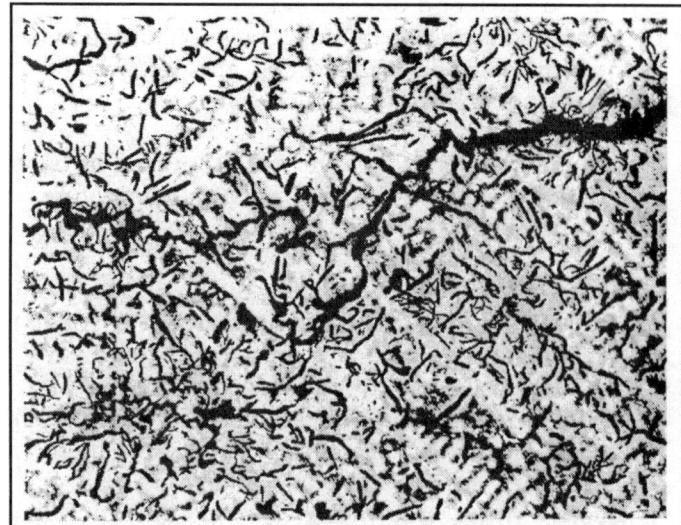

Figure 4. The Crack Tip Progressing Through a Link-Up of Graphite Flakes- 50X, 2% Nital Etch (Courtesy of Commonwealth Edison)

The crack tip was in a state of residual tension at room temperature. This was evident when the crack opened up as the crack-containing sample was removed from the head casting. The overall conclusion of the metallurgical evaluation was that the crack most likely resulted from thermal fatigue. The examination indicated that the crack probably initiated at the small radius where the flame face intersects the seat insert counterbore.

ENGINE TESTING - Engine testing was performed to validate the finite element model and to examine the sensitivity of cylinder head operating temperatures to reduced coolant flow and part load engine operation. Temperatures were continuously measured and recorded for a series of tests which were approximately one hour in duration. Comprehensive documentation of the test procedure and results is included in [8].

A KSV engine was made available for testing at Sumner Municipal Light and Power (SMLP) Station, a municipal power plant. The test engine had a lower rating and speed than engines used in nuclear stand-by service.

The cylinder head was instrumented with thermocouples near the center of all four valve bridges as shown in Figure 5. In addition, temperature sensitive "temp" plugs were installed on the flame face side of the cylinder head as shown in Figure 6. A pressure sensor was fitted to the cylinder head to record cylinder pressure at 1/16° crank angle increments. Temperatures were recorded during the transient start-up events and during continuous engine operation. Temperatures were measured at three-quarters and full load.

The sensitivity of cylinder head operating temperatures to a coolant flow restriction was examined to assess the importance of casting variation at the coolant outlet passage (Figure 3). A flow meter and flow control valve was installed downstream of the cylinder head coolant outlet passage. Temperatures were measured for full and reduced coolant flow rates of 222 and 62 liters/minute per cylinder, respectively. Measured temperature data indicated that the cylinder head operating temperatures were fairly insensitive to changes in the coolant flow rate. Exhaust valve bridge temperatures only increased by about 15°C when the coolant flow rate was reduced by 72%.

STRUCTURAL ANALYSIS - A thermal-mechanical finite element analysis was performed to predict cylinder head operating temperatures and stresses for power generation applications. The model was used to analyze cylinder head response for five operating conditions listed in Table 1.

The engine cycle simulation program WAVE/IRIS [9] was matched to the engine performance data measured at the SMLP Station. Measured and predicted cylinder pressures for 100% of rated power

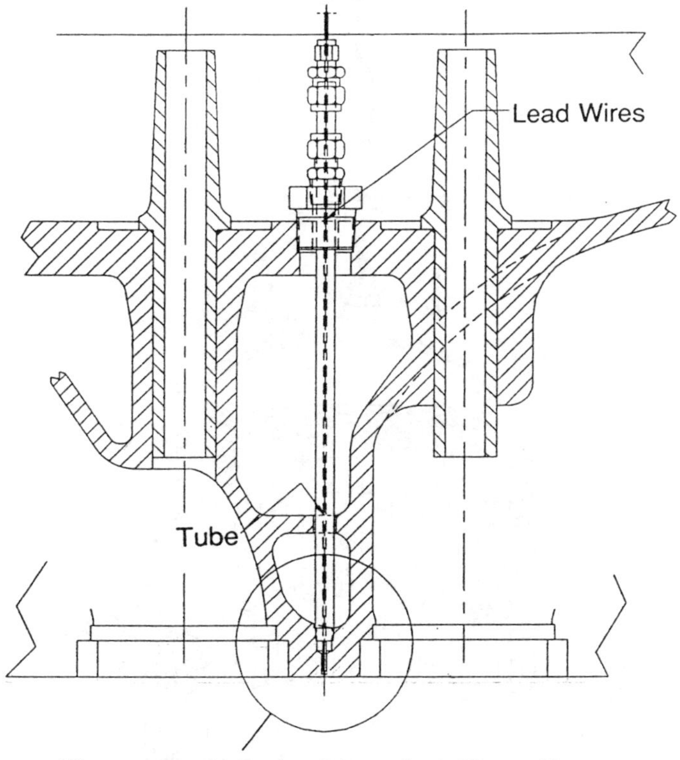

Figure 5. Thermocouple Installation at the Exhaust-Intake Valve Bridge

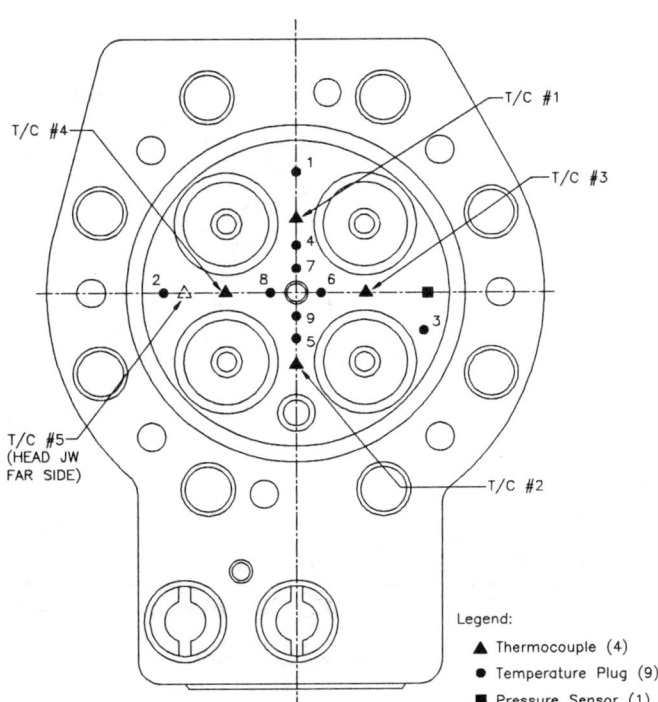

Figure 6. Thermocouple, Temperature Plug and Pressure Sensor Locations

Table 1: KSV Engine Operating Conditions

Station	Percent of Rated Power	Engine Speed [RPM]	Brake Power [kW/cyl]	BMEP [bar]	P max [bar]
Sumner	75%	514	175	10.55	87
Sumner	100%	514	230	13.86	105.1
Sumner	110%	514	252	15.18	108.8
Braidwood	100%	600	274	14.13	116.1
Braidwood	110%	600	300	15.51	120.6

are compared in Figure 7. WAVE/IRIS calculated spatially resolved convective film coefficients and bulk gas temperatures in-cylinder [10], which were subsequently used as boundary conditions for a steady state heat transfer analysis. A full transient analysis of the engine unloading condition was not necessary to meet the objectives of the investigation.

A 3-D solid finite element model was generated with sufficient detail to properly simulate cylinder head interaction with other major engine components, including the valves, valve seat inserts and liner (Figure 8). A nominal fire deck thickness was used for the initial analysis. Since the engine has unit heads that are nearly symmetric, a half-model was sufficient and allowed a finer mesh density.

A steady state heat transfer analysis was performed to predict cylinder head temperatures for constant load engine operation. A temperature correlation exercise was used to validate the modeling approach and gain further confidence in the analysis results. Predicted temperatures were within 10% of thermocouple measurements for full and three-quarters rated power at the SMLP Station (Table 2). The measured temperatures were mean averaged values from a series of one hour tests. Consistently higher temperatures were measured at thermocouple T/C #3 due to a flow blockage through that valve bridge caused by the pressure sensor. Predicted temperatures were generally within 10% of the temp plug measurements.

Predicted temperatures were highest locally around the rim of the seat insert counterbores (Figure 9a). Temperatures tended to be highest near the narrowest part of the valve bridges. Operating temperatures at the exhaust valve bridge are compared for two different engine speeds in Figure 10. The variation of cylinder head temperatures was approximately linear with respect to BMEP for a given operating speed. Temperatures increased at the higher

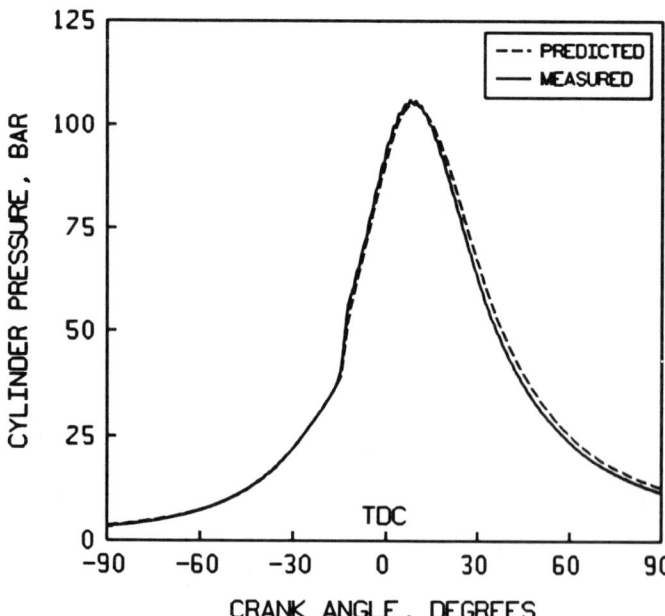

Figure 7. Cylinder Pressure Correlation for a KSV Engine Running at 100% of Rated Power at the Sumner Municipal Light Plant

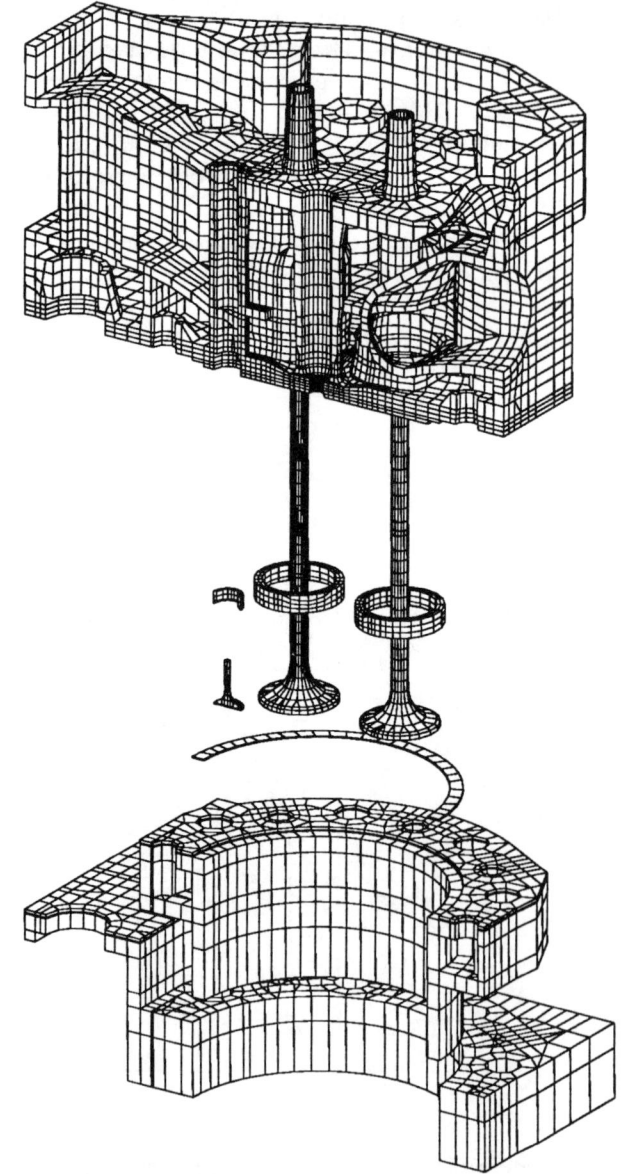

Figure 8. 3-D Finite Element Model of the KSV Cylinder Head Assembly

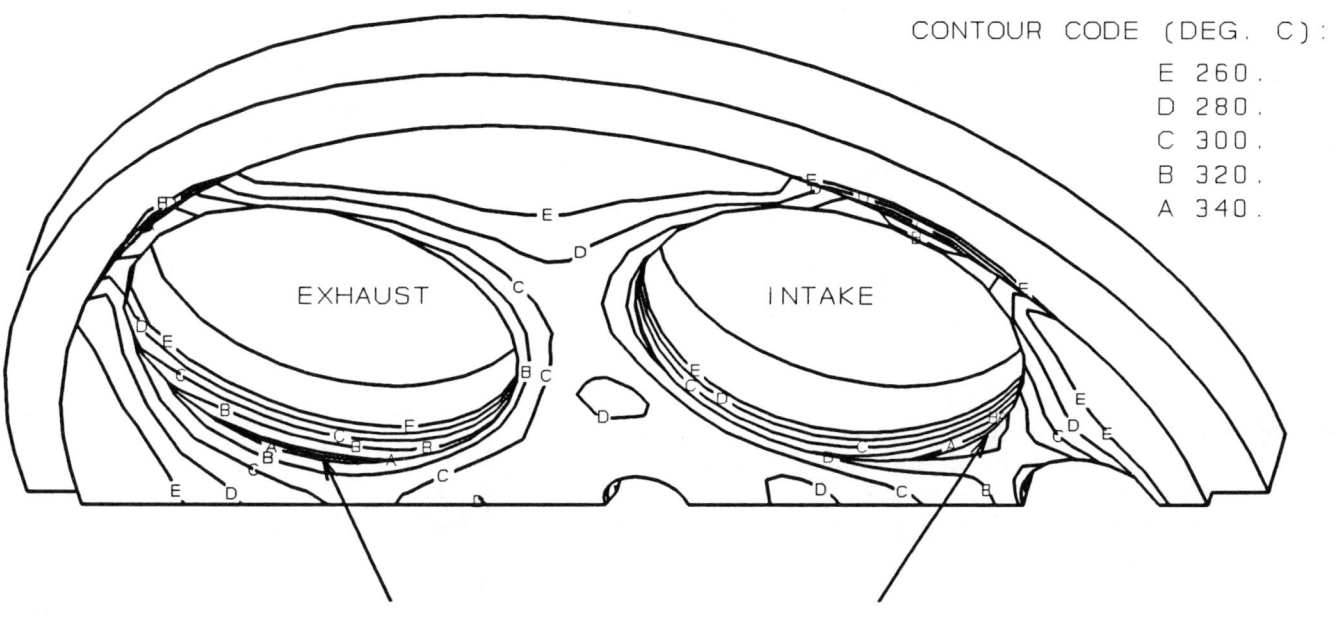

Figure 9a. Predicted Temperatures on the Flame Face of the Cylinder Head for 10% Overpower Engine Operation at the Braidwood Station

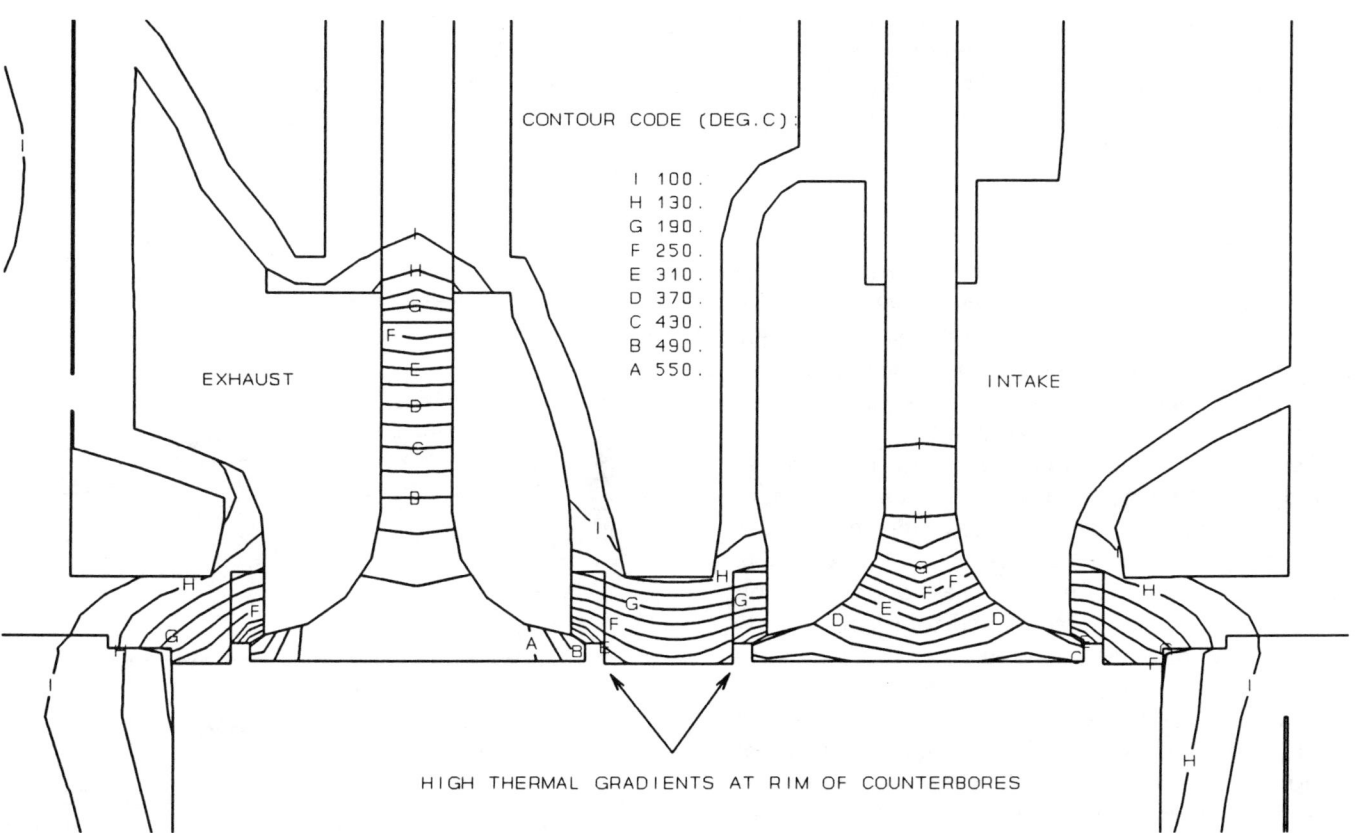

Figure 9b. Predicted Temperatures on a Cross-Section Through the Exhaust and Intake Valves for 10% Overpower Engine Operation at the Braidwood Station

Table 2: Thermocouple Temperature Correlation

Operating Condition	T/C Location	Measured Temp (°C)	Predicted Temp (°C)	Percent Difference
Sumner 100% Power Full Coolant Flow	1	245	257	4.9
	2	239	248	3.8
	3	266	262	-1.5
	4	239	262	9.6
Sumner 100% Power Reduced Coolant Flow	1	260	274	5.4
	2	251	264	5.2
	3	287	278	-3.1
	4	251	278	10.8
Sumner 75% Power Full Coolant Flow	1	226	234	3.5
	2	234	226	-3.4
	3	251	240	-4.4
	4	237	240	1.3

speed because more fuel energy is released per unit of time. The peak temperature at the exhaust valve bridge was 355°C for 10% overpower engine operation at the Braidwood Station. This predicted temperature falls below the threshold where cracking problems typically occur for gray iron cylinder heads (<380°C, [2]).

High localized temperature gradients were found at the small radius around the seat insert counterbores (Figure 9b), and cause large compressive hoop stresses during engine operation. The thermal expansion around the hot rim of the counterbores is constrained by the lesser expansion of the bottom deck, causing compressive hoop stresses. The compressive state of stress at the valve bridges is primarily induced by thermal loading, but increases by 15 to 20% when the cylinder fires near TDC due to peak cylinder pressure loading. The peak compressive stress was at the narrowest part of the exhaust valve bridge. Analysis of the exhaust valve bridge region was of most interest, since it was the highest stressed location and coincided with the primary crack initiation site reported from the inspections and engine owner surveys. Moreover, the model was in general agreement with the short radial cracks seen in service (the secondary cracking mode), as high compressive hoop stresses were predicted around the perimeter of the seat insert counterbores.

Operating stresses are compared at the exhaust valve bridge for two different engine speeds in Figure 10. The large compressive stresses were of interest for assessing structural durability. Compressive stresses may cause permanent plastic deformation locally in the valve bridge area, and the onset of low cycle thermal fatigue damage when the thermal load is reduced. Yield factors (the ratio of compressive yield strength to stress) were calculated to compare the potential for exhaust valve bridge yielding at different operating conditions. The yield factor is not an absolute measure of low cycle fatigue durability, but rather a comparative parameter to assess the relative potential for material yielding. Yielding would promote more severe compressive-tensile hysteresis loading. The yield factor varied considerably for the engine loads and speeds under consideration, confirming that

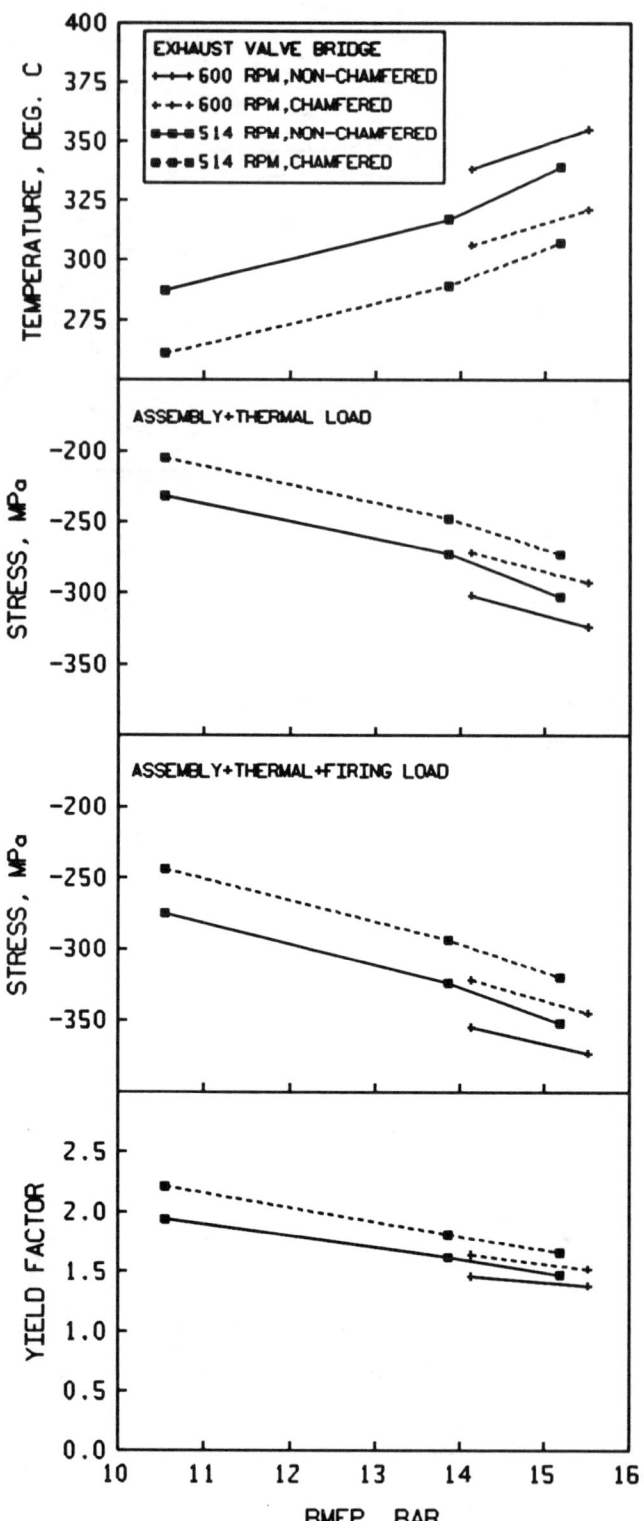

Figure 10. Exhaust Valve Bridge Response for the Current Design (Solid Line) and Chamfered Design (Dashed Line) at Five Engine Operating Conditions

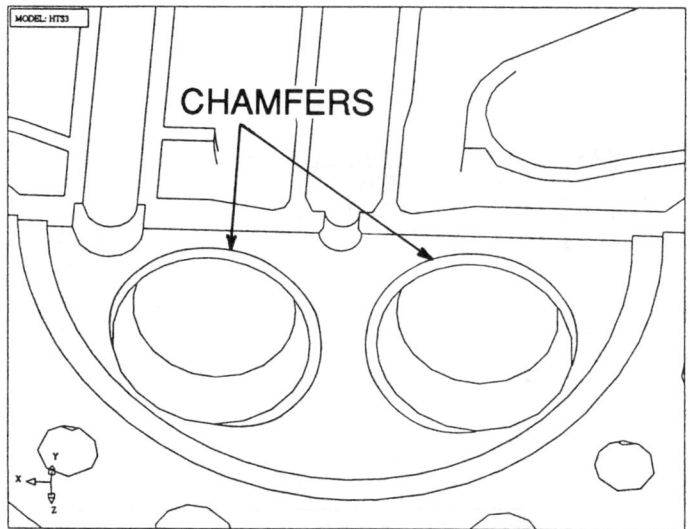

Figure 11. Crescent-Shaped Chamfers Proposed Around the Valve Seat Insert Counterbores

cylinder head durability is sensitive to engine rating.

Valve seat insert press-fit and head bolt-up loads were included in the analysis. The interference fit between the seat insert and counterbore was found to be appropriate. Small tensile hoop stresses develop in the valve bridges due to cold press-fit of the inserts, but assembly stresses were not large enough to cause bridge cracking.

SENSITIVITY STUDIES - The initial analysis indicated that the nominal cylinder head design was structurally sound, with the exception of the high temperature gradients and stresses around the seat insert counterbores that induce permanent yielding and may eventually cause crack initiation. It was proposed that small machining changes, such as chamfers, scallops, flycuts or reduced deck thicknesses would be sufficient to reduce the localized stresses to improve the service life of the cylinder head. Two machined features were examined.

<u>Counterbore Chamfers</u> - Crescent-shaped chamfers were proposed around the valve seat insert counterbores to eliminate the small-radiused rim responsible for high thermal gradients (Figure 11). A full chamfer could not be machined around the circumference of the counterbore, as it would interfere with the combustion sealing gasket.

The analysis was repeated using a sub-modeling approach to evaluate the effectiveness of the chamfered seat insert counterbores. The chamfers effectively reduced operating temperatures, thermal gradients and stresses around the counterbores (Figure 12). Peak temperatures and stresses at the exhaust valve bridge are compared with the current non-chamfered design in Figure 10. A 10% reduction in operating temperatures and stresses was achieved.

Detailed fatigue test data was not available for

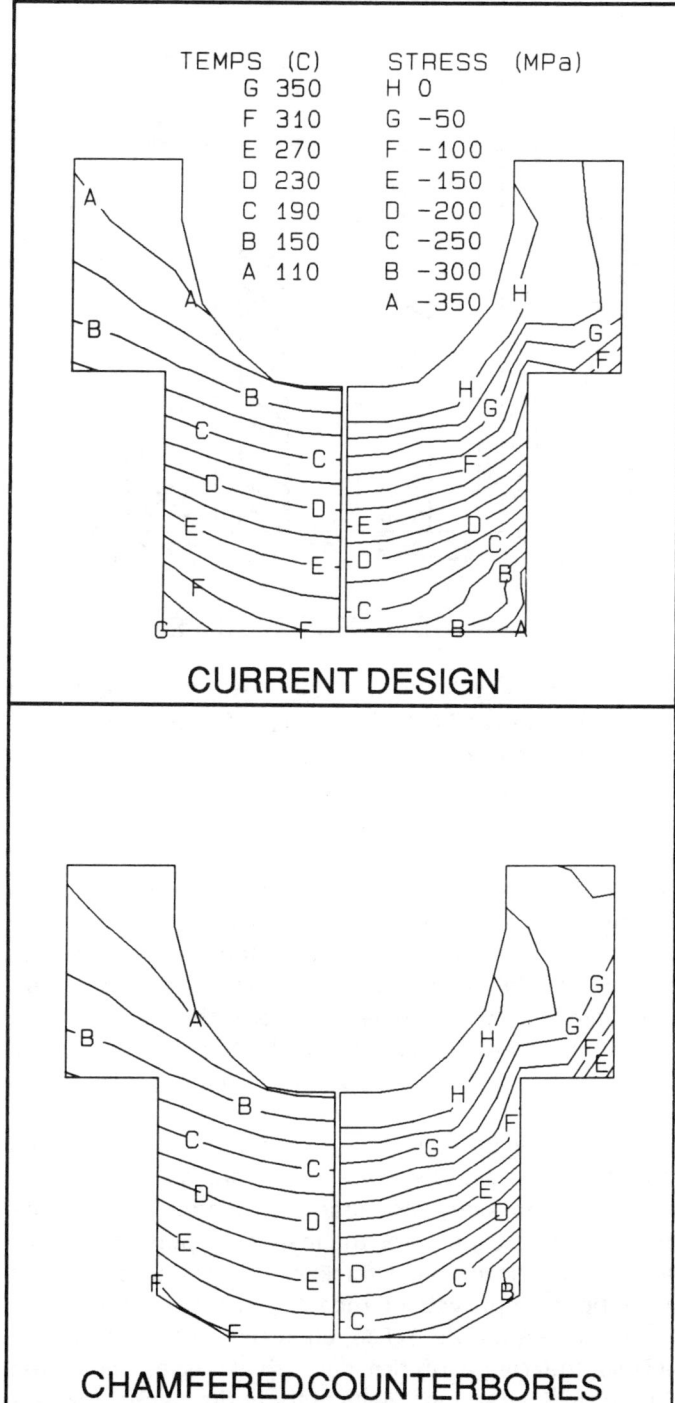

Figure 12. A Comparison of Exhaust Valve Bridge Temperatures (Left Side) and Stressess (Right Side, Assembly+Thermal+Firing Load) for 10% Overpower Engine Operation at the Braidwood Station

the specific alloy under consideration, so absolute fatigue life predictions were not possible. However, data for a similar gray iron permitted prediction of relative changes in fatigue life. The improvement in service life was estimated using strain-controlled fatigue test data at elevated temperatures for a similar

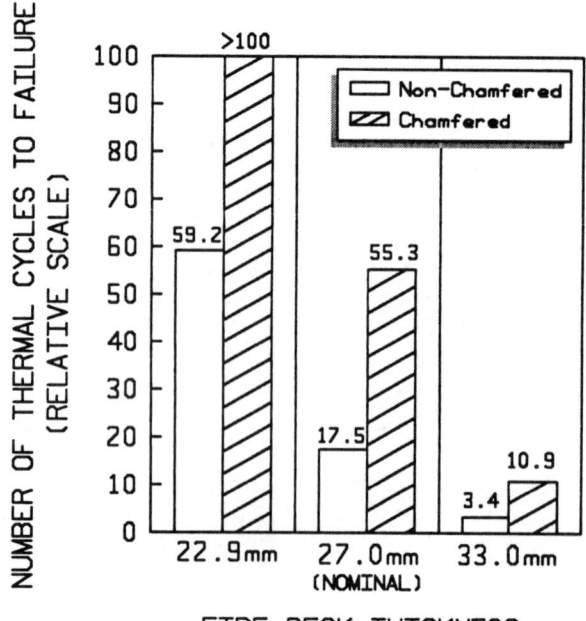

Figure 13. Predicted Fatigue Life at the Exhaust Valve Bridge for 10% Overpower Engine Operation at the Braidwood Station

cast iron material. Test specimens were loaded in compression and returned to zero *engineering* strain to replicate compressive-tensile hysteresis loading that occurs at the valve bridge. The chamfers were predicted to extend the service life of the cylinder head by about three times for 10% overpower engine operation at the Braidwood Station (Figure 13).

Fire Deck Thickness - Cylinder head inspections revealed considerable variation in fire deck thickness for a large population of cracked cylinder heads. The mean deck thickness for the inspected population was larger than the nominal thickness. Casting processes have been improved in recent years to reduce deck thickness variation. The analysis was repeated to assess the range of fire deck thicknesses measured during the inspection. Sensitivity to fire deck thickness was examined in order to estimate the variation in service life for the previous generation of heads, and to estimate the potential gains that will be realized by reducing deck thickness variation via improved casting processes.

Exhaust valve bridge temperatures and stresses are compared for three different fire deck thicknesses in Figure 14. Stress contours for the thickest fire deck are compared in Figure 15. Predicted fatigue life varied by a factor of 17 for the range of deck thicknesses measured during inspection (Figure 13). This variability is considered excessive and may have lead to increased failure rates in service. The efforts taken to reduce fire deck thickness variability should significantly improve the service life of new production cylinder heads.

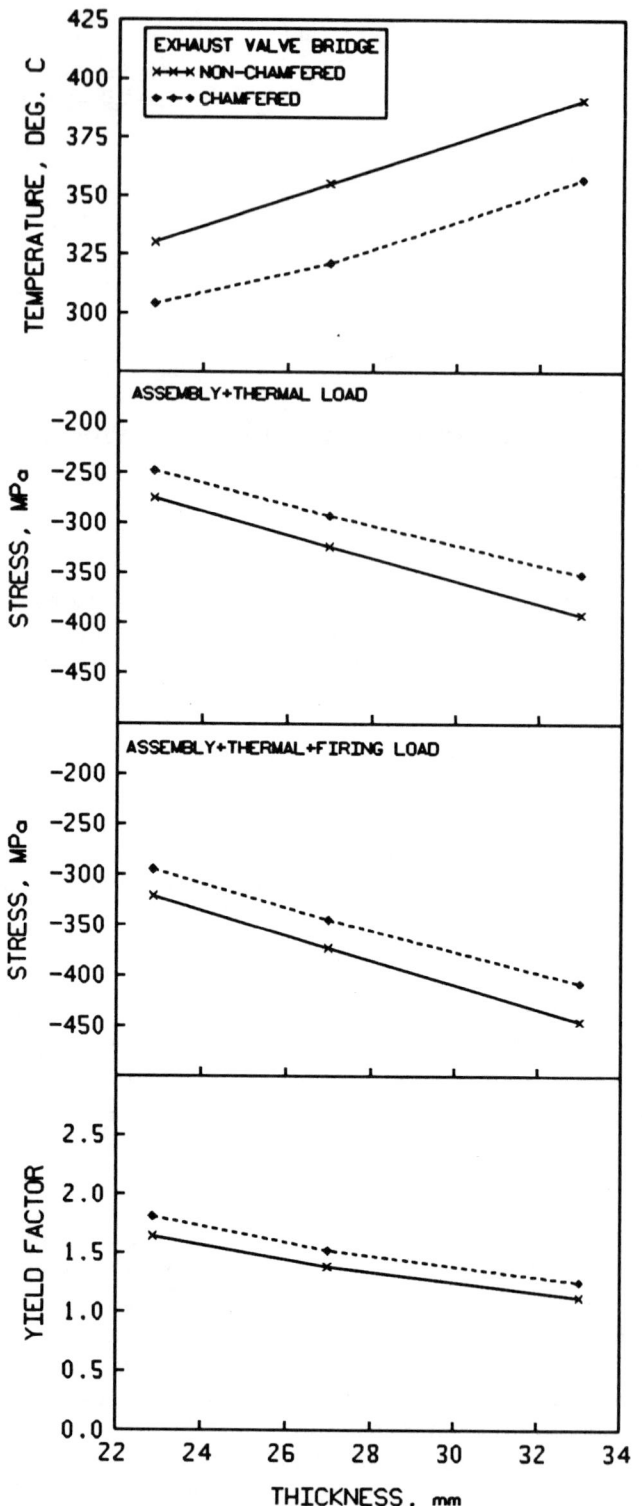

Figure 14. Sensitivity of Exhaust Valve Bridge Temperatures and Stresses to Fire Deck Thickness for 10% Overpower Engine Operating Conditions at the Braidwood Station

Moreover, analysis results indicated that a thinner fire deck design may be possible to further extend the fatigue life of the cylinder head. The combination of chamfered seat insert counterbores and a thinner deck design would greatly improve the durability of the cylinder heads.

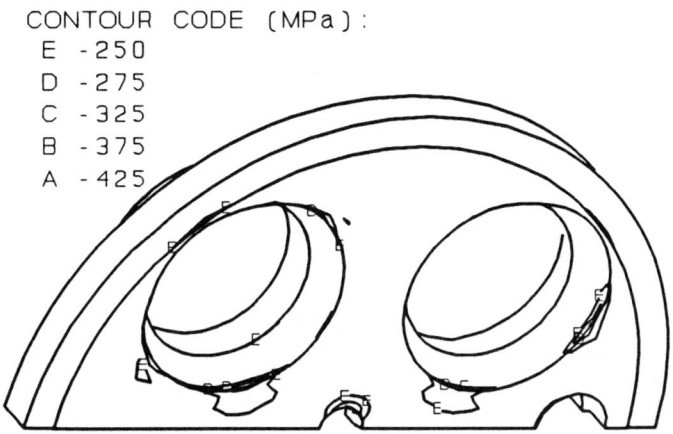

(A) 22.9mm FIRE DECK, CHAMFERED

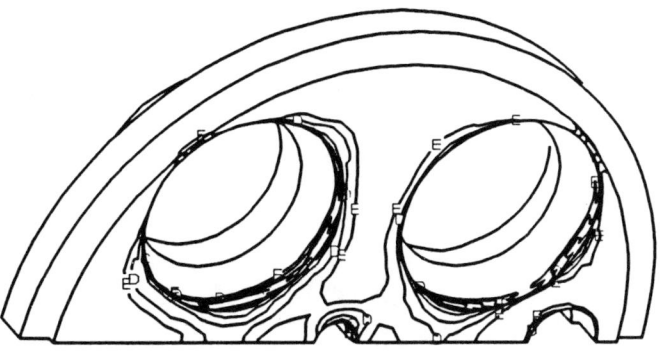

(B) 27.0mm FIRE DECK, CURRENT DESIGN

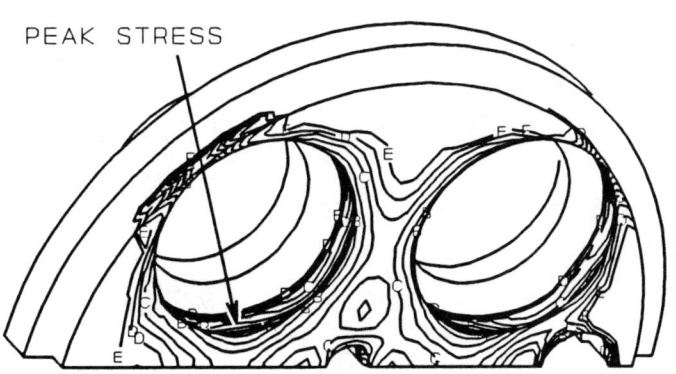

(C) 33.0mm FIRE DECK, EXTREME CASE

Figure 15. A Comparison of Cylinder Head Stresses Caused by Combined Assembly, Thermal and Cylinder Firing Loads for 10% Overpower Engine Operation at the Braidwood Station

CONCLUSIONS

The root cause investigation produced numerous conclusions and recommendations. The following conclusions are among the most significant.

1. The valve bridge cracking failure mode is primarily driven by thermal loads which produce high compressive stresses during continuous engine operation. This may cause more severe compressive-tensile hysteresis loading and produce low cycle thermal fatigue damage.

2. Analysis results indicate that the counterbore chamfers are an effective design change and would improve the service life of the KSV cylinder head by a factor of three for nuclear standby service. The proposed chamfer design would not require casting or seat insert design changes, and could potentially be used to retrofit or refurbish cylinder heads in service.

3. The cylinder head operating temperatures were insensitive to reduced coolant flow rates. Improved dimensional control of the coolant outlet passage or higher coolant flow rates are not warranted.

4. The casting process improvements taken to reduce fire deck thickness variation should significantly reduce the failure rate of new production cylinder heads.

5. The analysis confirmed that the cylinder head durability is sensitive to the engine rating.

6. The variation of cylinder head temperatures was approximately linear with respect to BMEP for a fixed operating speed.

7. A steady state finite element analysis was sufficient to assess the structural durability of the cylinder head design. The critically stressed locations identified from the analysis coincided with crack initiation locations seen in service. In this case, a transient analysis was not warranted since appropriate corrective actions could be identified from the steady state analysis.

ACKNOWLEDGEMENTS

The authors are grateful to the following individuals who assisted in the root cause investigation: John Horne, Cooper Industries; the Cooper-Bessemer Owner's Group; Terry O'Brien, Neil Mares, Dave Pederson, Terry Van de Voort and Brian Wilson, Commonwealth Edison; Bob Jergens, Jim Duhrkopf, Tim Duhrkopf and Wendell Bohle, Sumner Municipal Light Plant; Art Killinger and Craig Mauch, MPR Associates; Mete Alpan and Michael Nilsson, Ricardo North America; Chris LeLeux, Preventive Maintenance Systems.

REFERENCES

[1] Haller,C.L. et.al., "Evaluation of Cracking in Cooper-Bessemer Model KSV Diesel Engine Power Heads", MPR-1367, with Appendices from Commonwealth Edison and Ricardo North America, 1993.

[2] French,C.C.J., "Problems Arising from the Water Cooling of Engine Components", Proc. IMechE, London, Vol. 184, Part 1, No.29, 1969/70.

[3] Shalev,M., Zvivin,Y., and Stotter,A., "Experimental and Analytical Investigation of Heat Transfer and Thermal Stresses in a Cylinder Head of a Diesel Engine", Int. J. Mech. Sci., Vol. 25, No. 7, 1983.

[4] Smith,L.W.L., Angus,H.T. and Lamb,A.D., "Cracking in Cast Iron Diesel Engine Cylinder Heads", Proc. IMechE, Vols. 1 & 5.58, 1970/71.

[5] Helmich,M.J., and Ulrey,C.S., "Design and Development of KSV Engine", Trans. of the ASME: Journal of Engineering for Power, 1963.

[6] Blech,J.J., "Heat Transfer and Stress Analyses of Engine Heads and Evaluation of Methods of Preventing Head Cracks", SAE Technical Paper No. 820504, 1982.

[7] Houston Light and Power, "Cracked Heads on ESF DG#23", Significant Problem Investigation Report No. H880200, 1988.

[8] Haller,C.L. et. al., "Cooper-Bessemer Owner's Group Report on Testing of Cooper-Bessemer Model KSV-12-GDT Engine in Support of 1994 Technical Tasks", MPR-1561, 1994.

[9] Morel,T., Keribar,R. and Blumberg,P., "A New Approach to Integrating Engine Performance and Component Design Analysis Through Simulation", SAE Technical Paper No. 880131, 1988.

[10] Morel, T. and Keribar,R., "A Model for Predicting Spatially Resolved Convective Heat Transfer in Bowl-In-Piston Combustion Chambers", SAE Technical Paper No. 850204, 1985.

[11] Nozue,Y., Satoh,H. and Umetani,S., "Thermal Stress and Strength Prediction of Diesel Engine Cylinder Head", SAE Technical Paper 830148, 1983.

950520
Predictive Analysis of Lube Oil Consumption for a Diesel Engine

Walter Zottin, Marcos Clemente, and José Manoel Martins Leite
Metal Leve S.A. Ind. e Com.

ABSTRACT

One of the main trends in diesel engine technology is the reduction and control of lubricant consumption, since the lube oil constitutes a major contributor to the particulates exhaust gas emissions. Many efforts have been recently dedicated to the development of oil consumption simulation tools. This paper presents a methodology based on piston and ring dynamics simulation, coupled with an additional code that simulates the engine oil consumption. An example of the procedure is shown for a typical DI diesel engine, where some piston design parameters influence are studied. Finally, some engine tests results are presented in order to validate the numerical prediction.

INTRODUCTION

In addition to the exhaust gas emissions regulations that require ever reducing levels of lubricant oil consumption, engine durability and reliability can be listed as a major driving force for modern diesel engine technology. As a means to achieve all these targets, reduced oil consumption has to be achieved and maintained for extended mileages, coupled with low levels of blow-by, since this is known to be a major source of corrosive combustion products and abrasive soot particles that are responsible for lubricant oil degradation and wear of engine components.

Competitiveness of the diesel truck market also requires reduced lead times between conception and launching of new engine models. Any simulation tool that contributes to the prediction of engine components structural and functional performance will help in saving development time and costs, by reducing required number of parts samples, engine and laboratory tests.

This paper presents an example of the application of state of the art predictive numerical codes to simulate blow-by and oil consumption of a modern heavy duty DI diesel engine. Five different piston design configurations are simulated in order to optimize piston pin off-set and ring zone profile for reduced oil consumption and blow-by. Validation engine tests results have been performed and correlation with numerical predictions are presented and discussed.

ENGINE SPECIFICATION - The main characteristics of the studied engine are shown below:

Fuel	:	Diesel
Injection Type	:	Direct
Number of Cylinders	:	6, In Line
Rated Power	:	187.5 kW @ 2200 rpm
Compression Ratio	:	17.3:1
Peak Combustion Pressure	:	13.2 MPa

SIMULATION DESCRIPTION

A set of programs has been used for the complete simulation of lubricant oil consumption and blow-by. Figure 1 schematically shows the programs and the phenomena that are considered in the simulations.

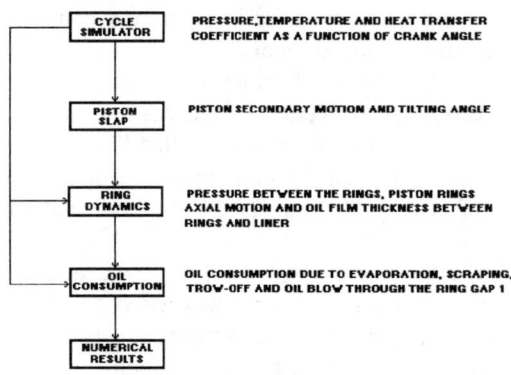

Figure 1 - Oil Consumption Numerical Simulation Flow Chart

As can be seen in figure 1, the results of piston and rings dynamics simulations are input data to a program that can simulate the engine oil consumption. The name of the latter program is OILCON [1] and was developed by AVL. The following oil consumption mechanisms are considered in this code:

- *Evaporation* : It is the part of the oil adherent to the cylinder liner that evaporates by the high gas temperature.
- *Scrapping* : It is the oil scrapped by the top land during the upward stroke and burned by the high gas temperature.
- *Blow-up* : It is the oil which flows through the end gap of the first ring into the combustion chamber where it is burned by the high gas temperature.
- *Throw-off* : It is the amount of oil above first ring which is exposed to the acceleration forces. The amount of oil above first ring is the sum of the oil scrapped by the first ring and the oil that flows at the side faces of first ring. A part of this amount of oil can be burned by the high gas temperatures.

PISTON SECONDARY MOTION - In order to simulate the piston secondary motion, it was used a program called PISDYN [4], which was developed by Ricardo North America, Inc. The mathematical model used in this program is based in a forces and moments balance applied to the piston. It also considers the elasto-hydrodynamic relationship between piston skirt and cylinder liner. Main output comprises piston's lateral and rocking motions, hydrodynamic oil pressure fields, skirt's elastic deformation, contact pressure between piston and liner (wear load).

RING DYNAMICS SIMPLIFICATIONS - The dynamics of the rings was simulated through a Metal Leve's proprietary program called RING [2]. The following simplifications were considered in this code :

- The piston moves up and down concentrically in relation to the cylinder liner.
- It was not considered the ring twist effect.
- In order to simulate oil film thickness between ring and cylinder liner, it was considered the fully flooded theory.

INPUT DATA - Piston's mass moment of inertia and position of its center of gravity were calculated through the CAD/CAE software called IDEAS.

Connecting rod mass moment of inertia and its center of gravity position have been estimated based on similar designs.

Cylinder liner distortion due to the assembly and thermal loads have been estimated based on Metal Leve's experience, and can be seen in figure 2.

Piston temperature distribution and piston displacement due to the thermal load were based on finite elements analysis.

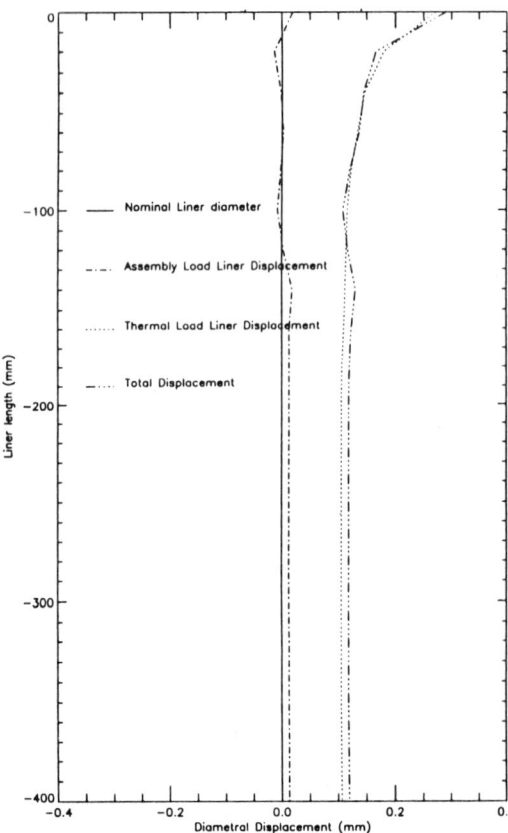

Figure 2 - Cylinder Liner Distortion Due To Assembly and Thermal Loads.

PISTON VERSIONS - The analyzed piston versions are the following:

- **Version 1** : Baseline piston.
- **Version 2** : Baseline piston plus a "J" groove at the land between first and second rings (second land).
- **Version 3** : Baseline piston plus a pin offset of 1.0 mm to the major thrust side.
- **Version 4** : Baseline piston plus a pin offset of 1.0 mm to the minor thrust side.
- **Version 5** : Baseline piston plus a pin offset of 1.0 mm to the minor thrust side and a "J" groove at second land.

THEORETICAL RESULTS

CYCLE SIMULATION - Results of the engine cycle simulation are shown in figure 3. Combustion pressure, gas temperatures and convective heat transfer coefficient are given as functions of the crank angle.

PISTON DYNAMICS - Figure 4 indicates the piston secondary movements during the engine cycle for all five simulated versions. This figure also includes the elastic deformations predicted for the piston skirt.

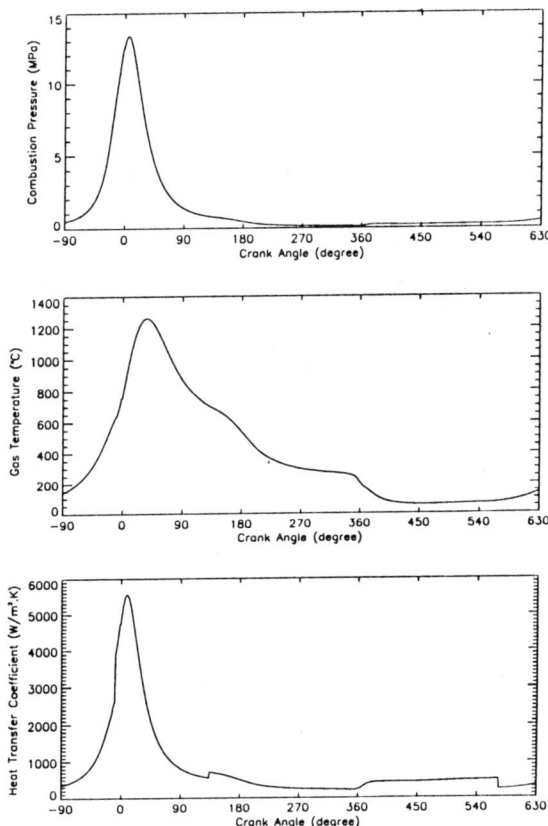

Figure 3 - Results of Engine Cycle Simulation.

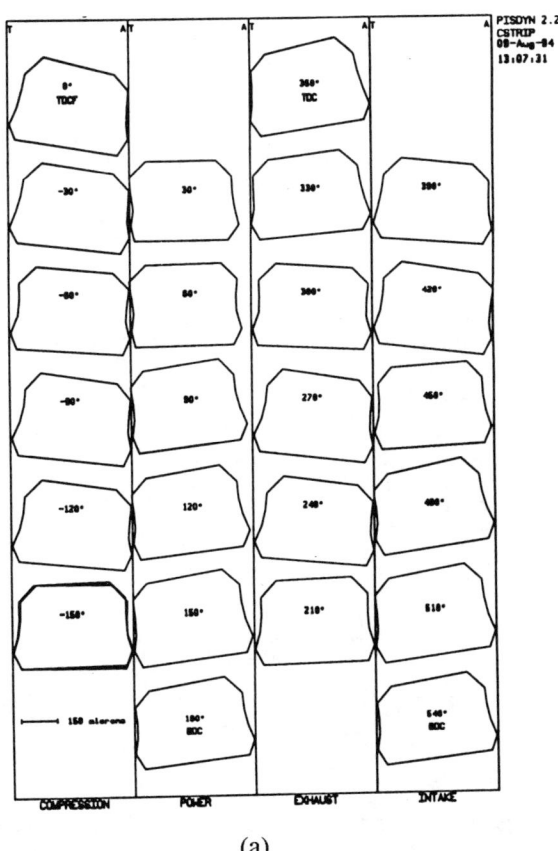

(a)

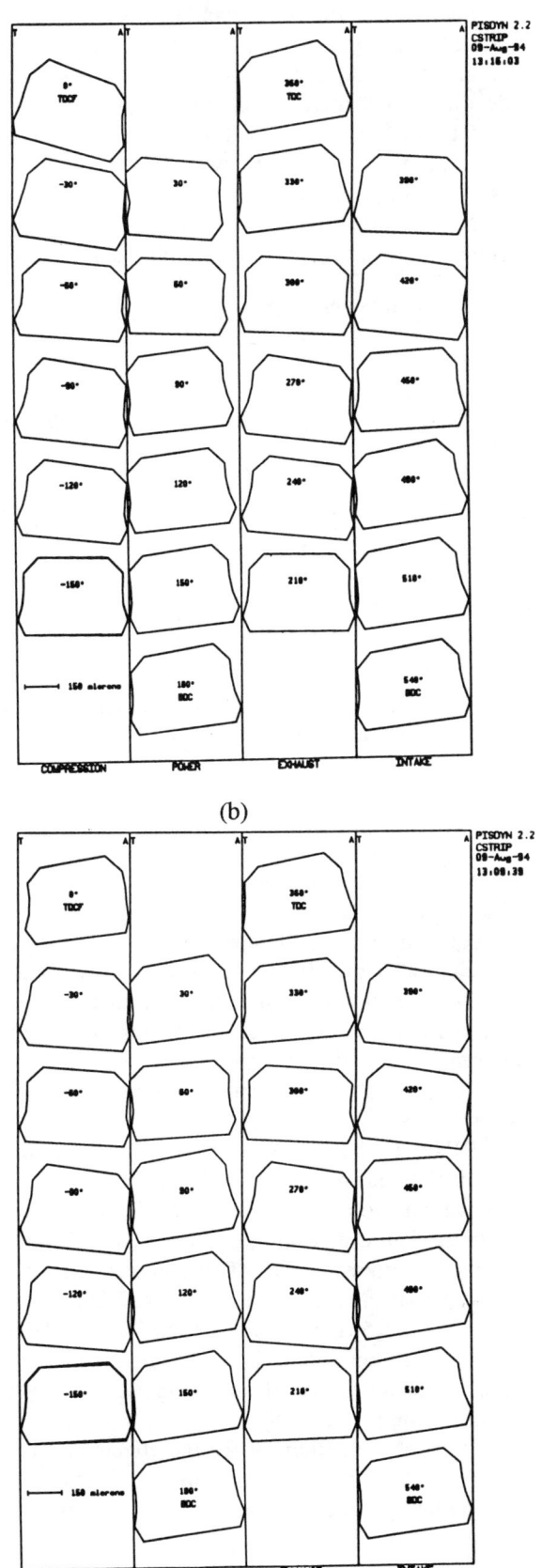

(b)

(c)

Figure 4 - Piston Secondary Movements and Skirt Elastic Deformations During Engine Cycle: (a) Versions 1 & 2; (b) Version 3; (c) Versions 4 & 5.

Piston's rocking angle can be seen in figure 5. The effect of piston pin-offset is clearly indicated. Version 3 (pin off-set to major thrust side) tilts to the minor thrust side after TDC firing. The opposite happens to versions 4 and 5 (pin off-set to minor thrust side). For reduced oil consumption, usually the lower the piston tilting, the lower oil consumption, due to the better guidance and support for the piston rings. Also the scrapping mechanism tend to prevail with higher piston tilting.

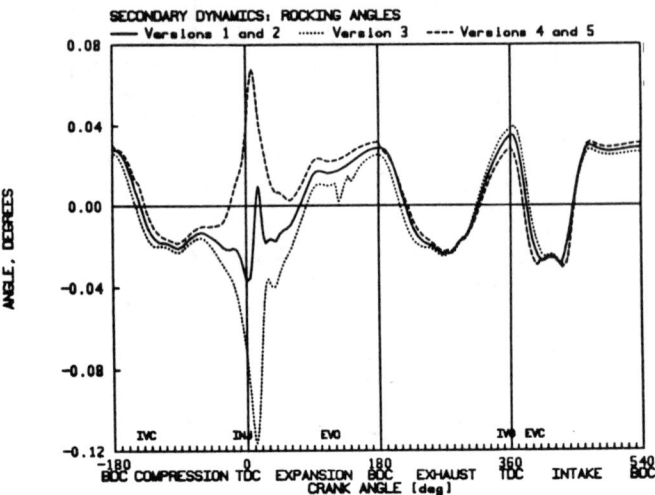

Figure 5 - Piston Rocking Angles.

Another functional affected by pin off-set is the hydrodynamic oil film between piston skirt and liner. At critical conditions this film may break or become so thin that boundary lubrication conditions may occurr. A wear load chart as shown in figure 6 is an output of the piston dynamics simulation. It indicates the areas of the piston skirt and position of cylinder liner most liable to present abnormal wear due to metal to metal contact, and indicate potential improvements of piston skirt profile and stiffness.

RING DYNAMICS - Pressure between first and second rings during engine cycle can be seen in figure 7. Relative position of the rings inside their grooves is shown in figure 8. Both figures clearly show the effect that the "J" groove between first and second rings has on pressure distribution in the ring zone. Versions with the groove (2 and 5) have lower pressure (figure 7), because of the larger volume for gas entrainment on second land, which probably reduces the lifting of the first ring around 90° crank angle (figure 8).

Illustrated in figure 9 are the effects on the engine instantaneous blow-by of the different pressure distributions among the rings due to the "J" groove.

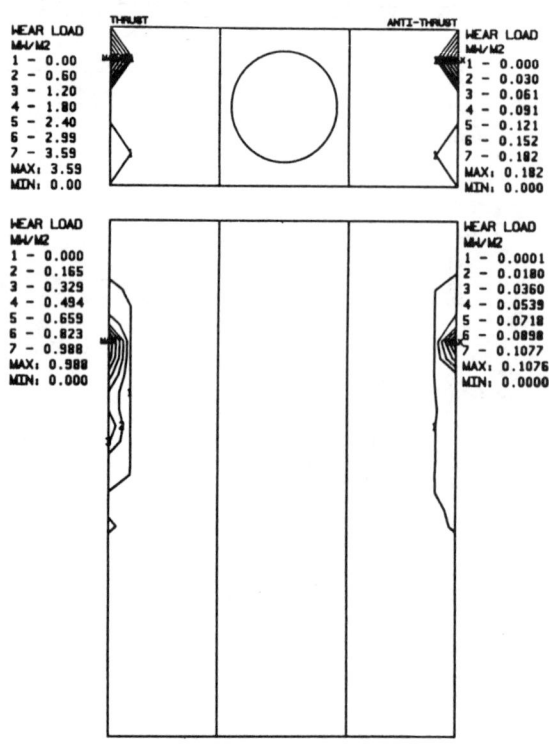

Figure 6 - Wear Loads On Cylinder Liner and Piston Skirt.

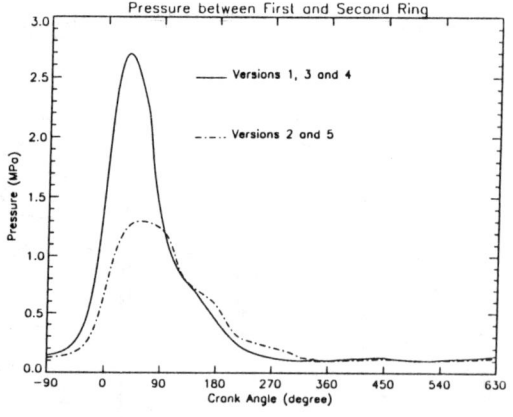

Figure 7 - Pressure Between First and Second Rings.

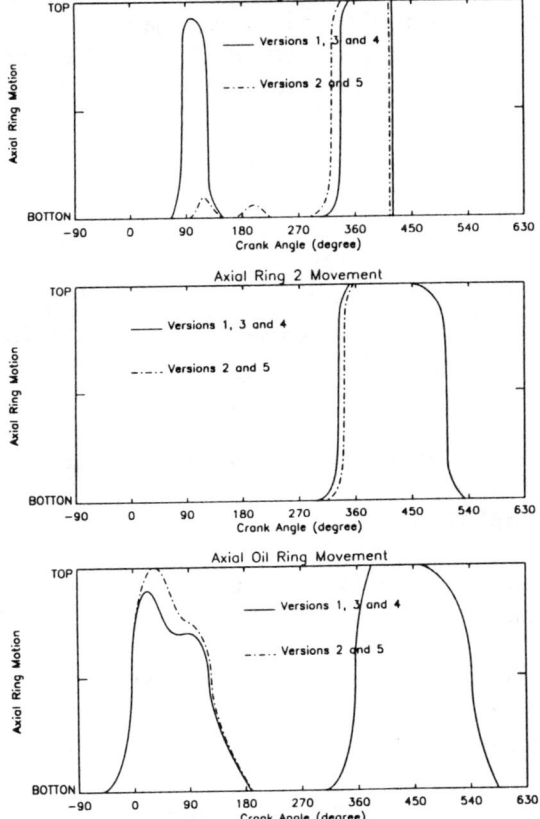

Figure 8 - Axial Movements of Piston Rings Inside Their Grooves.

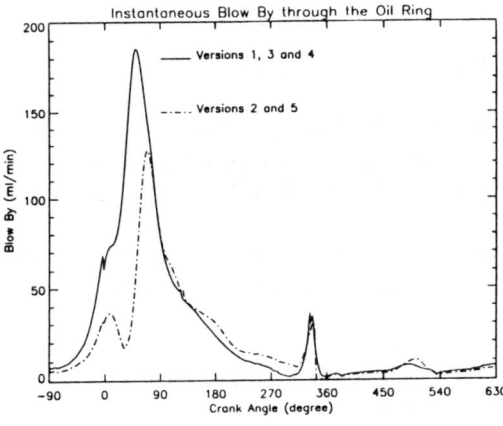

Figure 9 - Instantaneous Engine Blow-By.

OIL CONSUMPTION SIMULATION - Figure 10 represents the instantaneous lube oil consumption predictions for the five piston versions. The different oil consumption mechanisms are separated, and the effect of the pin off-set on the scrapping is clearly indicated, as is the "J" groove on the oil throw off mechanism.

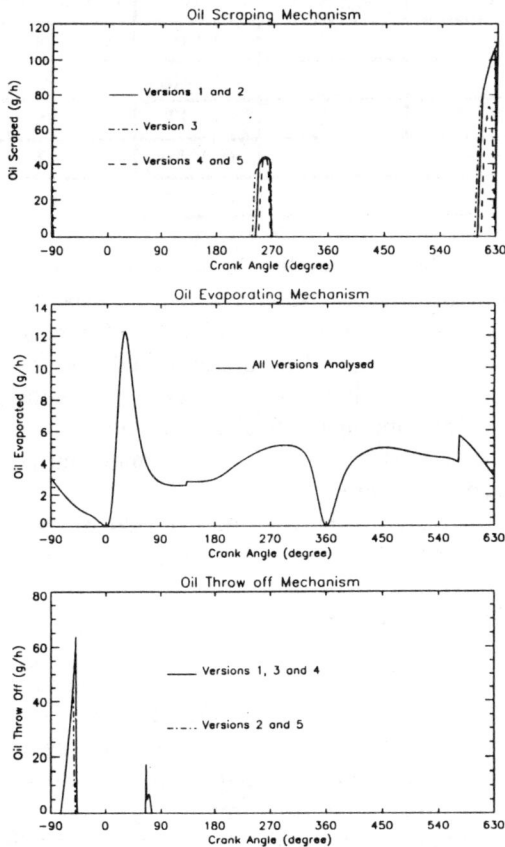

Figure 10 - Instantaneous Oil Consumption By the Different Mechanisms.

Table 1 shows the comparison of blow by and lubricant oil consumption for the five different piston versions.

Table 1 - Numerical Prediction of Blow By and Oil Consumption.

Version	Blow By (l/min)	Blow By (%)	LOC (g/kW-h)	LOC (%)
1	120.53	100.0	0.303	100.0
2	87.69	72.8	0.291	96.0
3	120.53	100.0	0.339	111.8
4	120.53	100.0	0.232	76.6
5	87.69	72.8	0.220	72.6

Table 2 shows the individual participation of different oil consumption mechanisms [1] which were considered in these analyses.

Table 2 - Oil Consumption Mechanisms Participation.

Version	Evaporation (g/kW-h)	Scrapping (g/kW-h)	Throw-Off (g/kW-h)	Total LOC (g/kW-h)
1	0.122	0.146	0.0348	0.303
2	0.122	0.146	0.0226	0.291
3	0.122	0.182	0.0355	0.339
4	0.122	0.0759	0.0342	0.232
5	0.122	0.0765	0.0217	0.220

Table 1 shows that the best result of oil consumption was obtained by **version 5** which corresponds to baseline piston profile plus a pin offset of 1.0 mm to the minor thrust side and a "J" groove at second land.

Table 2 shows that the **scrapping** is the prevailing oil consumption mechanism for versions 1 to 3. In the case of versions 4 and 5, the predominant mechanism is the **evaporation**.

Versions 2 and 5 tends to decrease the blow by of the engine.

ENGINE TESTS

In order to validate the numerical simulation some engine tests have been performed. The basic idea was to run the engine with some of the piston versions analyzed and to compare the theoretical oil consumption and blow by with the experimental results.

Figure 11 and 12 show respectively, the comparisons between theoretical and experimental results of blow by and oil consumption obtained for versions 1 and 2.

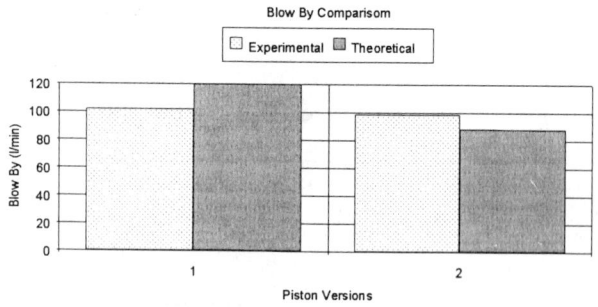

Figure 11 - Blow By Comparison

Engine tests show that the qualitative results obtained by the numerical simulations are correct. Quantitative differences observed between theoretical and experimental results may be reduced with further improvements of the numerical model, such as ring twisting, a better oil transport mechanism model and a better adjusting of some empirical coefficients.

Engine tests for versions 4 and 5 have not been performed yet but they are scheduled for the next year.

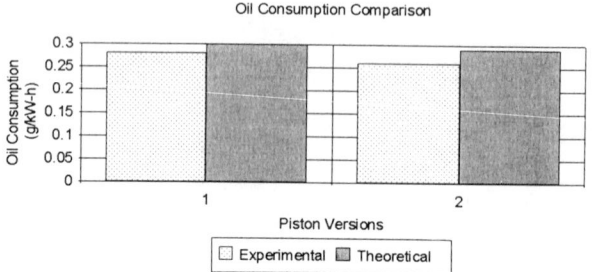

Figure 12 - Oil Consumption Comparison

SUMMARY

a) A procedure for the analysis of oil consumption numerical simulation has been described.

b) The proposed method was applied for a real diesel engine and the blow by and oil consumption were theoretically simulated for 5 different piston versions.

c) Some engine tests have been performed and a comparison between theoretical and numerical results was made.

d) Further improvements of the ring dynamics and oil consumption models will tend to reduce quantitative discrepancies between experimental and simulated results. Qualitatively, the numerical results are correct compared with the experimental results.

REFERENCES

[1] OILCON User's Manual, AVL List GmbH, 1992

[2] Zottin, W. Simulação da Dinâmica de Anéis de um Pistão utilizado em Motores de Combustão Interna, M.Eng. Thesis, POLI-USP, 1993.

[3] Duarte Jr, D.; Zottin W., Piston Slap Theoretical Manual, Technical Report # 383-008, Metal Leve S.A, 1993.

[4] PISDYN User's Manual, RICARDO NORTH AMERICA, 1993.

[5] Zottin, W.; Duarte Jr, D.; Perrone, L.F., Simulação Numérica da Dinâmica do Pacote de Anéis de Um Pistão

950521

Development of a New Engine Piston Incorporating Heat Pipe Cooling Technology

Qian Wang, Yiding Cao, and Alejandro Souto
Florida International Univ.

ABSTRACT

An increase in the temperature of charge in an engine combustion chamber is now more and more attractive due to its advantages in energy savings and environmental control. However, this will affect the design of engine elements, since a higher temperature will result in considerable thermal stresses and distortion, material and lubricant degradation, or even seizure and scuffing failures. Actually even for currently designed diesel engines, engine piston assembly failure, particularly piston ring/cylinder liner interface failure due to heat accumulation, is a very serious problem. Effective means of carrying heat away from this area are crucial for the prevention of scuffing. However, due to the reciprocal motion of the mechanism, efficient piston cooling is difficult to achieve using conventional cooling methods.

A newly developed engine piston incorporating heat pipe cooling technology aims at efficiently transferring excessive heat from the piston ring region to a location where cooling is sufficient, so that the temperature in the ring region can be reduced. Heat pipes can be implanted close to the ring grooves and move together with the piston assembly. This design utilizes the high heat conductance of heat pipes due to the two-phase heat transfer to achieve this heat transfer process, and employs the piston reciprocating motion as the means for the liquid phase return. Therefore, impingement of fluid droplets inside the heat pipe is critical for the new piston design. Dynamic analysis has been conducted and the results indicate that for a given fluid and pipe combination, the motion of the liquid droplets is a function of the piston stroke, cranking speed, heat pipe length, as well as drag and viscous resistances. Experimental dynamic observations have also been performed on an engine/heat pipe apparatus constructed based on a single-stroke internal combustion engine. Full scale liquid impingement on the entire inner wall of the pipe can be achieved when the angular cranking frequency was around and above 6-7 Hertz.

NOMENCLATURE

A crank length

a working fluid acceleration, m/s^2
d distance of droplet displacement
F_v drag force, N
F_w wall friction force, N
L pipe length, m
l connecting rod length, m
M equivalent mass of the piston/pipe assembly, kg
m mass of a fluid "particle," kg
r crank length, and amplitude of pipe motion, m
t time, s
t_0 time when the fluid "particle" departs from the pipe bottom wall, s
u working fluid velocity, m/s
v pipe velocity, m/s
x pipe displacement, m
ω angular velocity of cranking, rad/s

SUBSCRIPTS AND SUPERSCRIPTS

a after collision
hp heat pipe
i initial
l liquid
o before collision
r relative motion

1. INTRODUCTION

An increase in the temperature of charge in an engine combustion chamber is now more and more attractive due to its advantages in energy savings and environmental control. However, this will affect the design of engine elements, since a higher temperature will result in considerable thermal stresses and distortion, material and lubricant degradation, or even seizure and scuffing failures. Piston scuffing in diesel engines, particularly the piston/piston-ring/cylinder-liner interface failure, is a serious problem that degrades engine performance and shortens engine life. Many causes, such as piston thermal deformation and distortion, lubricant coking, excessive wear and debris deposition, as well as ring sticking, have been proven to induce scuffing [Wong et al., 1993]. These causes are closely related to the high temperature distribution in the piston assembly, especially the piston ring area, due to the excessive combustion heat [Wong et al., 1993; Heywood 1988]. Among assembly elements, the piston is the most vulnerable since heat accumulates there due to the difficulty in accessing cooling sources. One of the working limits for the piston is the maximum temperature that a piston can sustain. The rapid fall-off of the mechanical properties of the commonly used Al-Si alloy at temperatures above $200^0 C$ is responsible for piston ring sticking and piston material transfer due to contact adhesion and wear. A high temperature in the piston also results in an increase in the designed clearance between the piston and cylinder liner, which causes noise and vibration due to piston slapping. Cast iron pistons have been proposed to solve some of the problems. However, the temperature of the cast iron piston can be higher, usually about 40^0 to $80^0 C$ than that of Al-Si alloy pistons [Heywood, 1988], and the coking and degradation of lubricant oil may become serious.

At higher working temperatures, the problems mentioned above will become more serious. Piston design with effective cooling is crucial in preventing failure and improving engine thermal efficiency and service life. However, due to the reciprocating motion of the mechanism, efficient piston cooling is difficult to achieve using conventional cooling methods, such as crankcase oil splashing and internal gallery circulation for heavy duty diesel engines, as described by Law and Day [1969], and Mihara and Lidoguchi [1992]. Cao and Wang [1994] developed a novel cooling method using shaking-up heat pipes. Shaking-up heat pipes, like conventional heat pipes [Cotter, 1965 and Cao and Faghri, 1992], have an evaporator at the heat input section and a condenser at the cooling section. Evaporation in the evaporator section and condensation in the condenser section result in very effective heat transfer path and a very high heat conductance. The liquid return from the condenser section to the evaporator section in a shaking-up heat pipe is achieved by means of the shaking-up action of the pipe, and because of that, the liquid phase motion is also an effective means of heat transfer. It has been proven, in a separate study, that the shaking-up heat pipe, like conventional ones, has a very high thermal conductance [Cao and Wang, 1995]. The shaking-up heat pipe has the ability to transfer the excessive heat in the high temperature region to a location where accessibility to the cooling source is much greater. Incorporating the shaking-up heat pipe into an engine piston can provide an effective way to reduce the temperature of the piston, or the piston assembly, where tribological contacts form, and to develop engine pistons for higher thermal efficiency and better performance of the engine piston.

The current work focuses on the development of a new engine piston incorporating shaking-up heat pipes for cooling the ring groove region. Some key issues for design considerations are discussed. A simulation apparatus has been constructed for experimental exploration of the feasibility and effectiveness of heat pipe cooling.

2. A PISTON DESIGN INCORPORATING SHAKING-UP HEAT PIPES

Temperature distribution and heat flow paths in an engine piston are now well known. Normally, heat flux is the highest in the center of the cylinder head, in the exhaust valve seat region, and to the center of the piston [Heywood, 1988]. Figure 1 presents a representative diagram of the temperature field of a typical diesel engine piston, where the highest temperature exists at the lip of the bowl and the center of the crown, and heat flows from the crown to the ring grooves and the piston wall. In order for the groove temperature to be reduced, heat flow along the highest temperature gradient needs to be controlled. Inserting heat pipes into the region close to the ring grooves may provide an alternative route for heat conduction, so that the heat transfer rate through the piston groove and piston wall can be significantly reduced. Figure 2 schematically shows the structure of an engine piston with implanted shaking-up heat pipes. A few heat pipes arranged parallel to the piston wall are inserted into the region close to the piston top ring grooves, extending down to, or passing through, the bottom region of the piston skirt, depending on the available space in the piston-crank shaft assembly. The top and the bottom portion of the heat pipe will serve respectively as the evaporator and the condenser. Heat pipes will move simultaneously with the engine piston, and the fluid inside will be shaken up and down so that two-phase heat transfer can be achieved. Due to their very high thermal conductance, these heat pipes transfer the excessive heat away from the top ring region to the area where cooling by the crank case oil is much more effective. Thus, the ring groove temperature can be substantially reduced. On the other hand, if the ring area of the new piston works at the same temperature as that of the ring area of a conventional piston, the new design allows the piston assembly to work at a much higher power load. Since no contact is made between the heat pipe and the lower piston skirt wall, the thermal and lubrication conditions in the piston skirt area will not be affected. Rather, the temperature there is expected to be reduced too, since the heat transfer rate from the piston ring region to the upper skirt wall is reduced.

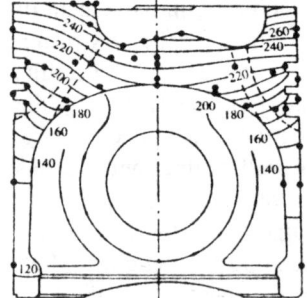 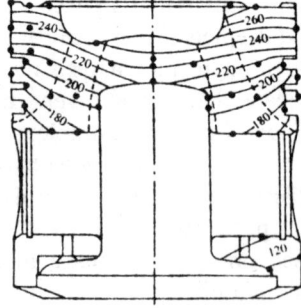

Figure 1: A representative diagram for isothermal contours (solid lines) and heat flow path (dashed lines) [Heywood, 1988]

Heat pipes can be manufactured from copper, aluminum, steel, or other metallic materials, and water can be employed as the working fluid. The cast-in method, or other mechanical and metallurgical methods, may be employed to install the heat pipes to the desired area. The heat pipe may have different configurations to fit the geometric space of the engine piston, and different end shapes to enhance the heat transfer areas, as shown in Fig. 3. A uniform distribution of heat pipe evaporators on in the upper section of the piston is expected, as shown in the cross section, A-A, in Fig. 2. However, due the piston pin and the connecting rod, heat pipes can occupy only a certain space down from the wrist pin bearings, as shown in the cross section, B-B, in Fig. 2. Therefore, not only straight but also curved heat pipes, as illustrated in Figs. 3 (a) and (b), are to be utilized. Heat pipes can also be designed with an enlarged stud-shaped end-cap, as suggested by Fig. 3 (c), a symmetric end-cap, and (d), a non-symmetric end-cap, to increase the heat transfer area between the heat pipe and the high temperature region of the piston and to reduce the distance between the adjacent heat pipes. The non-symmetric end-cap may be used to replace the curved heat pipe to meet the structural requirement. Fins can be attached to the condenser of the heat pipe, if necessary, to increase the cooling area.

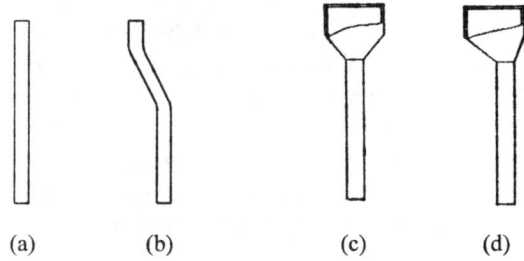

Figure 3: Schematic representation of the shaking-up heat pipes with different geometries.
(a) The straight heat pipe.
(b) The curved heat pipe.
(c) The symmetric stud-shaped heat pipe.
(d) The non-symmetric stud-shaped heat pipe.

3. FLUID IMPINGEMENT AND THE CORRELATION OF DESIGN PARAMETERS

This design employs the piston reciprocating motion as the means for the liquid phase return to achieve the high heat conductance of heat pipes. Therefore, impingement of fluid droplets inside the heat pipe is critical for the new piston design. A full scale droplet collision on the entire inner pipe wall, especially two end walls, is required for effective liquid/vapor transfer. A kinematic analysis was conducted first to shed some light on the fluid impingement and the fundamental relation of some geometric and dynamic parameters related to heat pipe and the piston design.

When the heat pipe is subjected to an oscillatory motion, either sinusoidally, or reciprocally as the motion of a piston of an internal combustion engine, the fluid inside the pipe may be subjected to a complicated motion and may exist in a mixture of fluid droplets and vapor. The droplet motion is of the utmost importance since it determines the success of the liquid return the heat pipe operation. As a first approximation, fluid droplets are simplified into "particles," and a kinematic analysis of a single particle is conducted, while the influence of vapor resistance can be treated as a drag force, and the interaction with the pipe wall may be considered by collisions. Influence due to "particle" interference is neglected in the current stage to simplify the analysis.

Liquid droplet impingement

Figure 4 shows the crank-slider mechanism and the coordinate system to describe the reciprocating motion. The heat pipe is represented by the close-ended cylinder attached to the slider, which symbolizes the engine piston. Displacement, x, of the slider can be expressed by the cranking angle, ωt, crank length, r, and the connecting rod length, l:

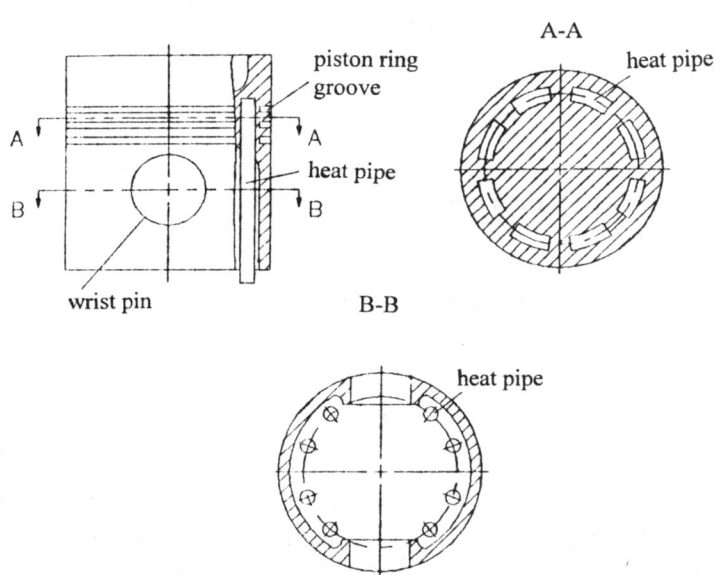

Figure 2: Schematic representation of the new piston design.

Incorporating heat pipes in piston design provides an effective method of heat transfer and a protection to the piston wall, especially the ring grooves region. Since the temperature there will be significantly reduced, the thermal distortion, friction, and wear will be well controlled. Although only the vertical piston assembly is illustrated and discussed in the current study, the same design principle applies to all kinds of piston arrangements.

$$x = A\cos\omega t + \sqrt{l^2 - A^2 \sin^2 \omega t} - l \quad . \quad (1)$$

The slider velocity and acceleration can be obtained from the time derivatives of Eq. 2 and are functions of $A\omega$, mechanism dimension, and the crank angular position. Figure 5 schematically shows the variation of the slider displacement, velocity, and acceleration with respect to the cranking angle. When the engine starts to crank, the fluid droplet "particle" stays at the lowest position of the pipe, as shown Fig. 4. The total force, ΣF acting on a "particle" of mass m may be the summation of gravity, resistance due to vapor drag, F_v, and wall friction, F_w, and the inertia force due to pipe motion:

$$\Sigma F = mg + (F_v + F_w) + m(d^2x/dt^2) \ . \qquad (2)$$

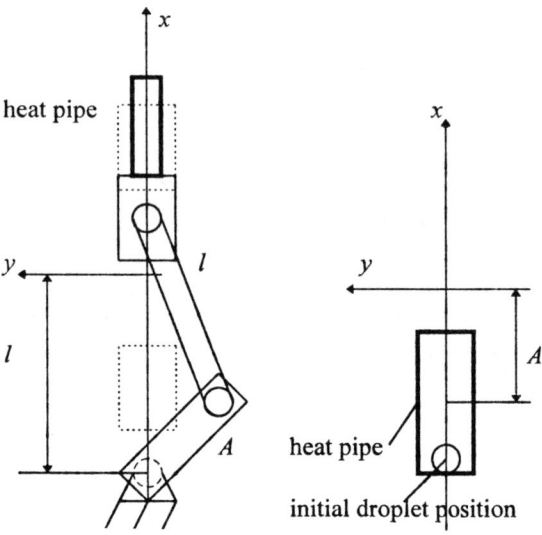

Figure 4: The crank-slider mechanism with a simulated heat pipe and the coordinate system arrangement.

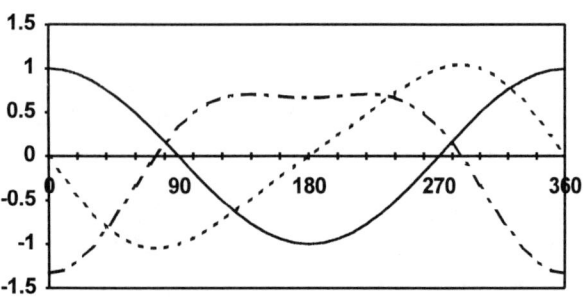

Figure 5: The position (x/A, solid line), velocity ($\dot{x}/A\omega$, dot line), and acceleration ($\ddot{x}/A\omega^2$, dash-dot line) of the piston slider and heat pipes.

The "particle" departs from the bottom wall at t_0 when ΣF becomes zero due to the direction change of the inertia force, $m\ddot{x}(t) = m(d^2x/dt^2)$, in the first half of this cranking cycle, as indicated by the slider/pipe acceleration shown in Fig. 5. Greater damping and wall resistance can result in later "particle"-wall separation and smaller initial speed, $u_{l,i}$, since the maximum $u_{l,i}$ occurs at the t_0 corresponding to zero pipe acceleration, $d^2x/dt^2=0$. The initial velocity of the "particle," $u_{l,i}$, which is the same as the pipe velocity at that moment when it leaves the pipe wall at time t_0 is given as follows:

$$u_{l,i} = \dot{x}(t) = v_{hp}(t_0) = dx(t_0)/dt \qquad (3)$$

Assuming a rectilinear motion of the "particles" inside the pipe, forces acting on a "particle" are expressed in the following equation, where a is for the "particle" acceleration:

$$mg + F_v - ma = 0 \ . \qquad (4)$$

The "particle" speed, u_l, is related to the initial speed, $u_{l,i}$, and the resistance, F_r, by:

$$u_l = u_{l,i} - \int_{t_0}^{t}(g + \frac{F_r}{m})dt \ . \qquad (5)$$

When the particle hits the end wall, collision may occur and the "particle" velocity is controlled by the collision process. Since the interaction time is considerably short, the momentum input due to the instantaneous speed variation of the piston may be neglected, and the following equation [Hibbeler 1989] can be applied to approximate the speed of the "particle" after such collision:

$$u_{l,i}^a = u_{l,i}^o - (1+k)\frac{M}{M+m}(u_{l,i}^o - v_{hp}) \ , \qquad (6)$$

where superscripts a and o are for the "particle" velocities after and before the collision, and M is the equivalent mass of the piston-heat pipe assembly. Parameter, k, is determined by the collision characteristics, being 1 and 0 for pure elastic and plastic collisions, respectively. Since m is much smaller than M, Eq. (6) can be simplified into the following form:

$$u_{l,i}^a = u_{l,i}^o - (1+k)(u_{l,i}^o - v_{hp}) \ . \qquad (7)$$

The collision parameter, k, may take some value between 1 and 0, and the real motion of the "particle" inside the pipe may be inbetween the motions of the "particles" subject to plastic and elastic collisions. The upper boundary of the "particle" velocity due to collision corresponds to a full elastic collision. The "particle" has a good chance of being accelerated by the pipe end walls, especially when a head-on collision occurs. The plastic collision provides the lower boundary for the "particle" velocity. After a plastic collision, the "particle" will adhere on the end wall and move together with it until the pipe acceleration changes its direction again at the second half of the cranking cycle, as indicated by Fig. 5. This process repeats as adhesion-separation for every half cycle of cranking.

The initial motion of a "particle" and its collision against the pipe end walls have been simulated in an ideal situation without considering the effects of resistance and gravity, and the results are presented in Fig. 6 for two different combinations of pipe and crank length. End wall positions are presented by the displacement with respect to the angular position of the crank. The "particle" initially moves together with the bottom end wall, and the first separation occurs at position, A, as marked on Fig. 6, at the crank angular position corresponding to a zero acceleration for the piston and the heat pipe under the assumed ideal condition. The "particle" then moves inside the pipe and hits the top end wall at position, B, where collision occurs either elastically, as presented by the broken line, e, or plastically, as presented by the broken line, p. The plastic collision is a repetitive process, while in the elastic collision process, the "particle" may bounce back and forth between the pipe end walls. It is evident that as long as the fluid "particle" can reach the top end wall after it leaves the bottom end wall, a full scale liquid impingement can be maintained, even if the collision was fully plastic. Figure 6 also indicates that the initial velocity of the "particle," the size of the heat pipe, and the length of the piston stroke control the collision position. With higher "particle" velocity, longer strokes, or shorter pipe lengths, collisions may be closer to the extreme positions of the piston stroke, as indicated by the apexes of the position curves of the end walls in Fig. 6.

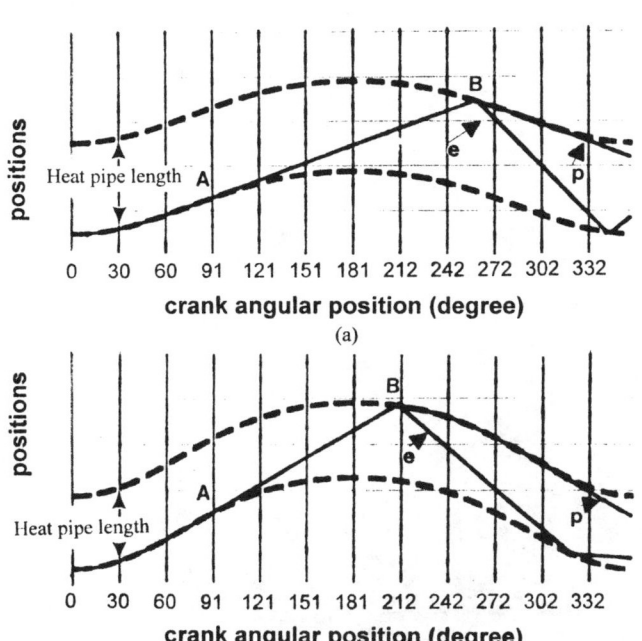

Figure 6: Motion of the fluid droplet (solid line) inside the heat pipe at 7 Hz angular speed of cranking, where end wall positions are indicated by the dashed lines.
(a) Heat pipe length: 101.6 mm
    Crank length: 34.9 mm
    Connecting rod length: 104.8 mm
(b) Heat pipe length: 81.6 mm
    Crank length: 50.5 mm
    Connecting rod length: 104.8 mm

Relation of the design parameters

Whether or not the "particle" inside the heat pipe can reach the top end wall after it departs from the bottom end wall determines the feasibility of the fluid impingement, and therefore, the success of the heat pipe cooling. The relative displacement of the "particle" at time, t, inside the pipe is an integration of the relative speed of the "particle" t_0 the pipe over a time period, $t- t_0$:

$$d_l^r(t) = \int_{t_0}^{t}(u_l - v_{hp})dt$$

$$= u_{l,i}(t-t_0) - \frac{1}{2}(g+\frac{F_v}{m})(t-t_0)^2 - \int_{t_0}^{t} v_{hp} dt. \qquad (8)$$

A "particle" can reach the top wall if this relative displacement is at least the length of the heat pipe before the particle speed is reduced to zero due to deceleration shown by Eq. (5). According to this lower boundary of "particle" speed, the time interval can be obtained:

$$t- t_0 = u_{l,i}/(g+F_v/m) . \qquad (9)$$

Substituting Eq. (9) into Eq. (8) for $d_l^r(t) = L$ results in the following relation:

$$u_{l,i}^2 /2(g+F_v/m) - \int_{t_0}^{t} v_{hp} dt = L . \qquad (10)$$

Knowing that $u_{l,i}$ is related to the production of the amplitude of the slider reciprocal motion and the crank angular velocity, ω, or:

$$u_{l,i} = c_1 A\omega , \qquad (11)$$

where c_1 is a proportional coefficient related to the position where the "particle" departs from the end wall. The second term of Eq. (10) may be simplified by the mean value of v_{hp} in the time interval of $t-t_0$, which can be related to $A\omega$ by another coefficient, c_2. For a given pipe length and engine working conditions, c_1 and c_2 are constants. Therefore, Eq. (10) may be rearranged into the following form:

$$\frac{(\frac{c_1^2}{2}+c_1 c_2)}{(g+F_v/m)}(A\omega)^2 = L .$$

If $c = (\frac{c_1^2}{2} + c_1 c_2)$, and the gravity effect is further neglected, the above equation becomes:

$$c\, m\, (A\omega)^2 /F_v = L . \qquad (12)$$

Equation (12) manifests a relation of some important design parameters, such as the crank length, A, which is one-half the piston stroke, the engine cranking angular speed, fluid "particle" mass, m, the drag force, F_v, and the effective pipe length, L. A longer heat pipe requires a larger piston stroke and/or higher cranking frequency, larger particle mass, and weaker drag. For a given pipe length, a configuration of the crank-slider mechanism, and a combination of the pipe and

(a)

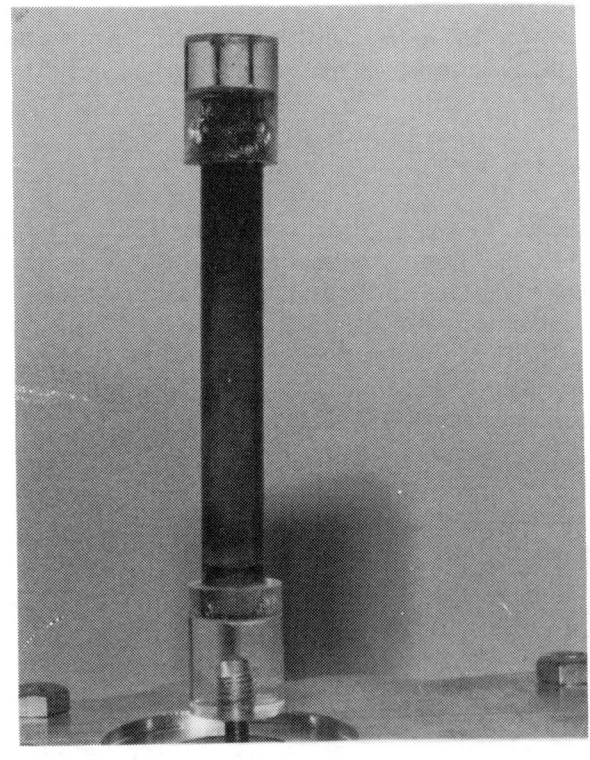

(b)

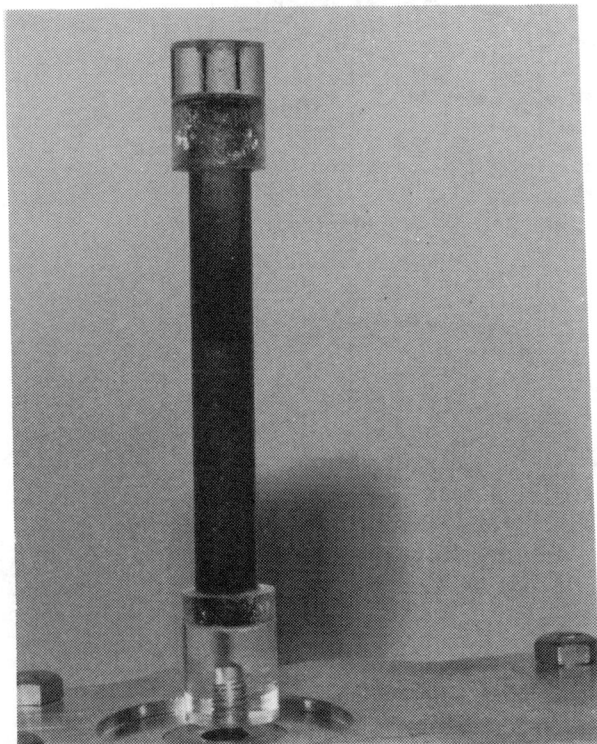

(c)

Figure 7: The experimental apparatus and photographs of the transparent pipe at 7 Hertz cranking motion.
(a) The experimental apparatus.
(b) Liquid impingement against the top wall.
(c) Fluid motion inside the pipe.

working fluids, there may exist a minimum cranking angular frequency, ω. On the other hand, the pipe length should balance the requirement for the minimum engine speed.

EXPERIMENTAL OBSERVATION OF THE FLUID IMPINGEMENT

A dynamic simulation apparatus has been designed and constructed as shown in Fig. 7(a). It utilizes the piston reciprocal motion of a single-cylinder internal combustion engine, Briggs & Stratton model 1904009. The piston stroke is 69.85 mm. The tested pipe is connected to the engine piston through a thread adapter so that it can experience the same motion as the piston due to combustion. A Balder electric motor, model M3559T, is used to drive the engine through a belt. The speed of the motor, which is also that of the engine crank, is controlled by a Toshiba motor controlling driver, model VFSX-2022p.

A 100 mm transparent pipe made from plexiglas with an inner diameter of 10 mm was filled with water colored with a green dye, which occupied about 10 to 15 percent of the interior volume of the pipe. The fluid stayed at the bottom position before the motor was turned on. When the pipe traveled with the engine piston, the fluid inside underwent an upright shaking motion, and was splashed on the wall of the pipe. The fluid splitting range with respect to the pipe increased as the cranking speed was increased. A full scale liquid impingement on the end-walls was observed when the cranking speed was around and above 6-7 Hertz, which can be referred to as a minimum impingement speed for the current setting. Figure 7 also presents a group of pictures taken at a 7 Hertz (420 rpm) engine motion. The liquid impingement against the top wall, as shown in Fig. 7 (b), and the liquid motion inside the pipe, as shown in Fig. 7(c), are clearly demonstrated. It is noted that this minimum impingement speed is well below the normal working speed of an engine, and it can be even lower if the piston stroke is longer, as for an actual automotive or heavy duty diesel engine. Since the vertical piston arrangement represents the worst impingement environment due to the greatest gravity effect, it can be conclude that a full scale liquid impingement can be established for any other engine pistons, too, without encountering technical obstacles.

Only part of a preliminary study of the piston design incorporating shaking-up heat pipes is presented in this paper. More work is planned toward full development of this technology and an optimized design of the engine piston.

CONCLUSIONS

A new engine piston has been designed employing the very high thermal conductance of shaking-up heat pipes to transmit the excessive heat in the top ring region to a location where accessibility to cooling oil is much greater, and the ring groove temperature is expected to be substantially reduced. A fundamental relation has been derived for some important design parameters, such as the piston stroke, which is twice the amplitude, A, the engine cranking angular frequency, ω, fluid "particle" mass, m, the drag force, F_v, and the effective pipe length, L. The initial motion and impingement of fluid droplets are analyzed. A full scale liquid impingement can be established and maintained by collisions, if the fluid can reach the top end wall in the first cranking cycle when the engine is started.

A simulation experiment has been conducted, and the minimum engine impingement speed, 6-7 Hertz, has been obtained for the current apparatus setting. A full scale liquid impingement can be established for engine application without encountering technical obstacles.

ACKNOWLEDGMENT: The authors wish to acknowledge Mr. Stan Vallidum and Ms. Helen Rooney at Florida International University for their help in machining the transparent pipe (Mr. Vallidum) and in preparing the manuscript (Ms. Rooney).

REFERENCES

Cao, Y. and Faghri, A. (1992), "Transient Multidimensional Analysis of Nonconvetional Heat Pipes with Uniform and Nonuniform Heat Distributions," Journal of Heat Transfer, Vol. 113, pp. 995-1002.

Cao, Y. and Wang, Q. (1994), " A New Engine Piston," Patent pending, Florida International University.

Cao, Y. and Wang, Q. (1995), "Piston Cooling and Shaking-up Heat Pipes and Performance Analysis of the Cooling System," to be presented at the 1995 SAE International Congress, Detroit, Michigan.

Cotter, T. P. (1965), *Heat Pipe Theory and Practice*, Hemisphere Publishing, Washington D. C..

Heywood, J. B. (1988), *Internal Combustion Engine*, McGraw-Hill Book Company.

Hibbeler, R. C. (1989), *Engineering Mechanics -- Dynamics*, 5th edition.

Law, D. A. and Day, R. A. (1969), "Oil Cooled Aluminum Alloy Diesel Engine Pistons--A New Approach," SAE paper No. 690749.

Mihara, K. and Kidoguchi, I. (1992), "Development of Nodular Cast Iron Pistons with Permanent Molding Process for High Speed Diesel Engines," SAE paper No. 921700.

Wong, T. Y., Scott, C. G. and Ripple, E. D. (1993), "Diesel Engine Piston Scuffing: A Preliminary Investigation," SAE paper No. 930687.

950522

Development of New Torsional Vibration Rubber Damper of Compression Type

Tomoaki Kodama, Yasuhiro Honda, and Katsuhiko Wakabayashi
Kokushikan Univ.

Shoichi Iwamoto
Saitama Univ.

ABSTRACT

The dynamic characteristics of the rubber dampers of compression type have been investigated in comparison with the conventional rubber dampers of shear type. The compression - type damper has been designed so as to produce compression force on the rubber part when torsional torque acts upon it.

This research report proposes the design method of the new compression - type rubber dampers.

The new rubber dampers have been fabricated on an experimental basis in accordance with the design method formulated by us. With the new dampers equipped in a 6 - cylinder, in - line diesel engine, the dynamic characteristics of stiffness and damping have been examined through experiments.

In comparison of the experimental results between the new compression - type rubber damper and the conventional shear - type rubber damper, it has been revealed that the compression - type rubber damper has some advantageous characteristics. The damping performance of the compression - type rubber damper is superior to that of the shear - type rubber damper particularly in the high - speed rotational range. Also, having excellent durability, lower possibility of variation in dynamic characteristics and smaller loss energy, the compression - type rubber damper is advantageous for torsional stiffness setting in damper designing.

INTRODUCTION

It is increasingly in demand to provide automobile diesel engines having higher specific power. To meet this requirement, the weight of diesel engine is reduced and the rigidity of cylinder block and crankshaft is decreased. In addition, the vibromotive force of torsional vibration is increased for higher output. Also, since the shaft parts of crankshafting are relatively thin with respect to moment of inertia at each mass point, significant low - order torsional vibration is generated in the driving rotational speed range. For reducing torsional vibration in the crankshafting, the shear - type torsional vibration rubber damper has been used in most diesel engines because of its inexpensiveness and compactness.

However, the shear - type damper may be damaged due to significant shearing strain produced in its rubber part, i.e. the durability and reliability of shear - type damper are rather low.

Taking it into consideration that the allowable compressive stress of rubber is larger than the allowable shearing stress, we have devised such a compression - type torsional vibration rubber damper as to catch torsional torque through rubber compression. In our experiment, the compression - type rubber damper has been made according to the design method formulated by us, and it has been mounted on the crankshafting of and actual engine to attain experimental result data of the characteristics of the stiffness and damping.

DESIGNING OF COMPRESSION - TYPE RUBBER DAMPER

Figure 1 shows the structure of the compression - type rubber damper that has been designed and fabricated in our study. As illustrated in this figure, the compression - type rubber damper is structured so that torsional torque is caught through compression of each rubber block sandwiched between blade parts in the housing and the inertia ring.

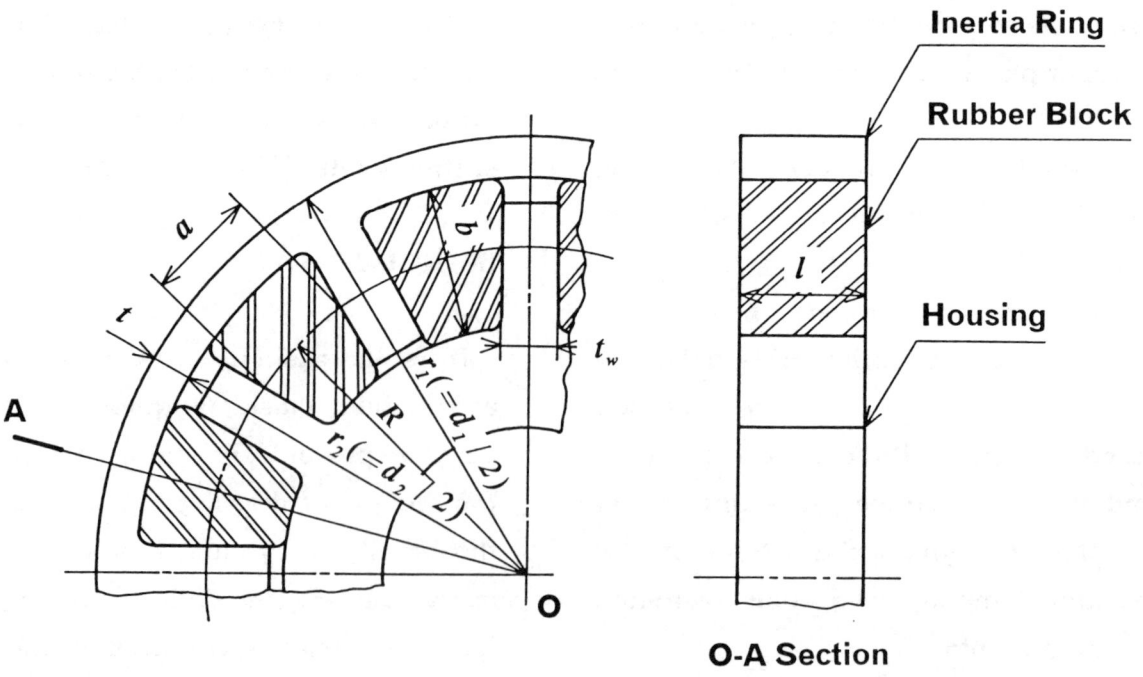

Figure 1: Shape of Torsional Vibration Rubber Damper of Compression Type

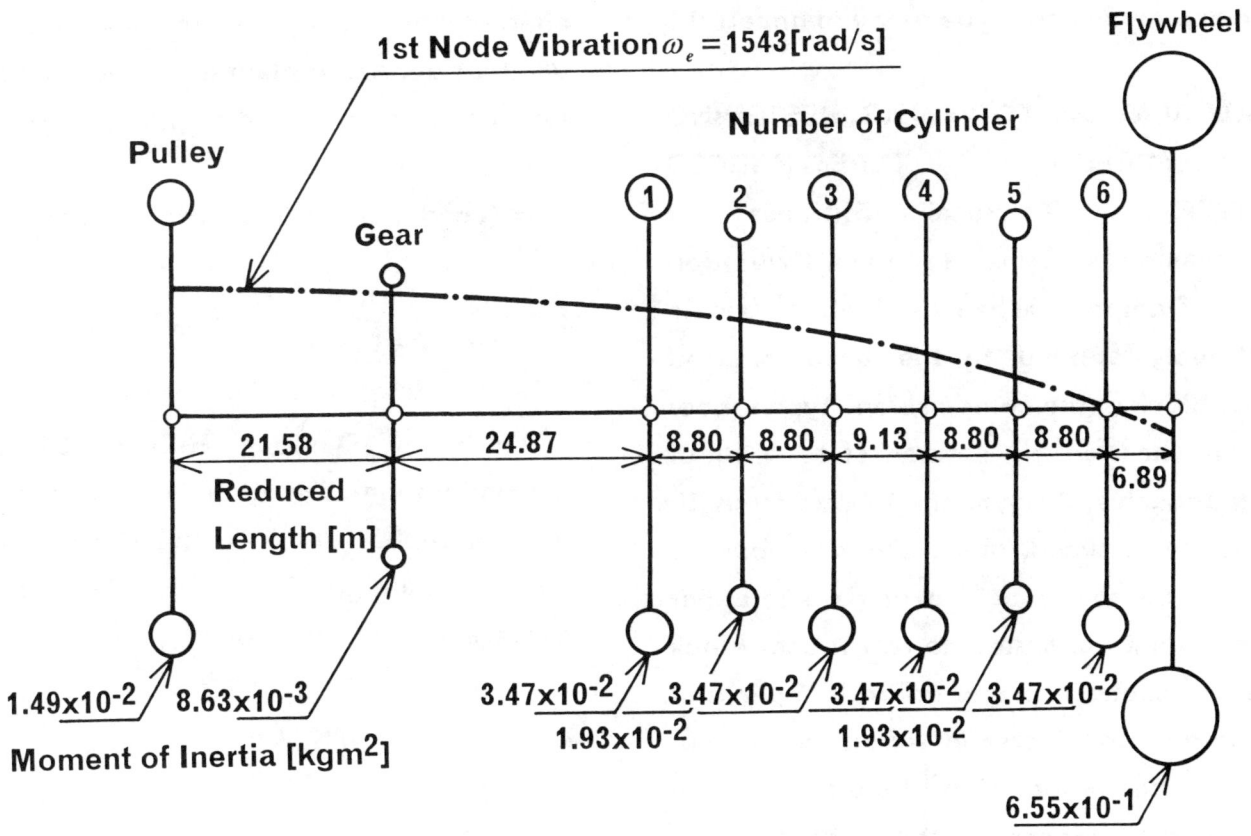

Figure 2: Torsional Vibration Equivalent System of Multi - Degree - of - Freedom of Test Engine

In designing of the compression - type rubber damper, it is important to presume accurate torsional stiffness of the damper in engine operation. For presuming torsional stiffness in engine operation, we have carried out simple dynamic vibration experiment to determine relationship between the spring constant and shape factor of rubber block. Figure 2 shows the equivalent torsional vibration mode of the engine under test, and Table 1 presents the major specifications of the engine under test. The compression - type rubber damper has been designed for the crankshafting of this test engine.

Table 1 Main Specifications of the Test Engine

Item		Contents
Main Use		Automobile Engine
Engine Form		In - Line
Cycle		4 Cycle
Number of Cylinders		6 Cylinder
Bore and Stroke	m	0.108 - 0.113
Total Stroke Volume	m^3	0.006211
Compression Ratio		18.9
Maximum Brake Output	kW/r/mim	112 / 3200
Maximum Brake Torque	Nm/r/min	402 / 2000
Firing Order		1-4-2-6-3-5

Described below is our design method taken for the compression - type rubber damper.

CALCULATION FORMULAS DESIGNING OF COMPRESSION - TYPE RUBBER DAMPER

Torsional Stiffness of Compression - Type Torsional Vibration Rubber Damper - When a vibration load is repetitively applied to the uncompressed rubber block of the compression - type rubber damper inserted in natural state without using adhesive, its dynamic fatigue strength is greatly reduced. Also, when a load is received by the front one of paired rubber blocks in natural state, the rear rubber block causes looseness in the entire damper unit, resulting in an adverse effect on the damper. Then, in the compression - type damper, we have forcibly inserted the rubber block having a rather smaller spatial size than that formed by the blades of damper inertia ring part and housing part. In this manner, an initial load has been applied to the rubber block.

Under the above initial load condition, the torsional stiffness of the damper can be expressed as shown below [1]*,[2]*.

$$K_d = \frac{2 \cdot n \cdot b \cdot l \cdot E_C \cdot R^2}{a} = 2 \cdot n \cdot R^2 \cdot K \qquad (1)$$

Shape Factor of Damper Rubber Part - Since the volume of rubber is not changed in non - compressive deformation of rubber, it is necessary to take account of the ratio of load area to free surface area (hereafter called " shape factor "). In case of the rubber block used for our experimental damper, the shape factor can be expressed as indicated below.

U = (Load area) / (Free surface area)

$$U = \frac{b \cdot l}{2 \cdot a \cdot (b + l)} \qquad (2)$$

Width of Damper Rubber Block in Circumferential Direction - The width of damper rubber block in the circumferential direction along the pitch line can be expressed as shown below.

$$a = \frac{\pi \cdot R}{n} - t_w \qquad (3)$$

In consideration of press - fitting of the rubber block, the allowable compressive strain is assumed to be 15 [%] to 20 [%]. Therefore, initial strain of 8 [%] has been applied in the compression - type rubber damper.

Under 8 [%] press - fitting condition, the size of the rubber block in natural state is represented as "$a_1 = a / 0.92$". Also, the length values of rubber block in the radial direction and axial direction can be determined using equations (1) and (2).

Accordingly, in determination of damper dimensions, it is required to take account of the relationship between compression spring constant and shape factor of rubber block, the relationship between torsional stiffness and inertia moment of damper inertia ring, and other restrictive conditions.

---

* Number in parentheses designate references at end of paper.

RELATIONSHIP BETWEEN COMPRESSION SPRING CONSTANT AND SHAPE FACTOR OF DAMPER RUBBER BLOCK

For designing of the damper, we have determined the compression spring constant of rubber block and the relationship between dynamic compression spring constant and shape factor of rubber block through free damping vibration experiments[3],[4].

The natural rubber pieces having Shore hardness values 50, 60 and 70 [Hs] (hereafter called " NR50", "NR60" and "NR70", respectively,) shown in Table 2 have been employed in the experiments. For each of these rubber pieces, four shape factor have been given.

Each rubber piece has been set in the press - fitting state, and the strain gauge has been applied in the compression direction. While hitting the rubber piece randomly with a hammer, we have recorded its compressive strain waveform on an electromagnetic oscillograph for determination of spring constant.

Table 2 Specifications of Test Rubber Block

Rubber Hardness and Material H_s	Shape Factor U	Static Spring Constant K_s N/m
NR 50	0.591	2.03×10^5
	0.422	1.21×10^5
	0.266	6.08×10^4
	0.206	4.71×10^4
NR 60	0.419	2.18×10^5
	0.408	2.28×10^5
	0.261	1.25×10^5
	0.203	1.02×10^5
NR 70	0.599	6.38×10^5
	0.422	3.49×10^5
	0.259	1.98×10^5
	0.203	1.93×10^5

NR:Natural Rubber

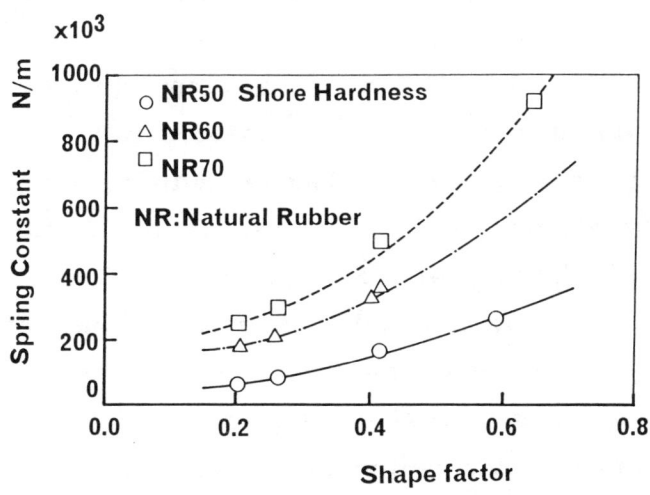

Figure 3: Relationship Between Shape Factor and Spring Constant (Rubber Block Temperature at 323 [K])

In this examination, the temperature of rubber piece has been changed in a range from 303 [K] to 363 [K] in increments of 20

[K]. In the result data of various engine experiments, it has been revealed that the temperature of rubber block of the damper mounted on the engine crankshafting is in a range of 313 [K] to 323 [K]. Figure 3 shows an example of relationship between shape factor and spring constant at a temperature 323 [K].

Note that the spring constant can be figured out by determining a natural frequency from the compressive strain waveform attained in the free damping vibration experiment.

RELATIONSHIP BETWEEN TORSIONAL STIFFNESS OF RUBBER DAMPER AND INERTIA MOMENT OF DAMPER INERTIA RING

According to reference [5]* and [6]*, the test engine crankshafting can be replaced by the two-degree-of-freedom equivalent vibration system that possesses two mass parts consisting of effective inertia of engine and inertia of damper inertia ring. In this vibration model, the optimum tuning ratio for the inertia moment of damper inertia ring [adjustment reference value for the equivalent vibration system of damper - less engine crankshafting] is expressed as shown below[5]*,[6]*.

$$\lambda_{opt} = \frac{1}{1+\alpha} \quad (4)$$

For meeting the compression-type rubber damper design requirements, substituting the following equations (5) - (8) into equation (4) yields the expression shown in equation (9).

$$\omega_d = \sqrt{K_d / I_d} \quad (5)$$

$$\omega_e = \sqrt{K_e / I_e} \quad (6)$$

$$\lambda_{opt} = \omega_d / \omega_e \quad (7)$$

$$\alpha = I_d / I_e \quad (8)$$

$$K_d \cdot I_d^2 + (2 \cdot K_d \cdot I_e - \omega_e^2 \cdot I_e^2) \cdot I_d + K_d \cdot I_e^2 = 0 \quad (9)$$

Since the natural frequency of the test engine and the effective inertia moment of the engine part are known already, equation (9) is applicable as the relational expression for the damper torsional stiffness and the inertia moment of damper inertia ring. Thus, using any given value of damper torsional stiffness, it is possible to determine an inertia moment value of damper inertia ring.

DETERMINATION OF DIMENSIONAL VALUES

After the examination of allowable mounting space in the engine, we have made the following assumptions [I] - [IV]. On these assumptions, we have determined dimensional values of the damper according to the aforementioned design procedures [4]*, [5]*.

[I] Thickness of blade part :

In consideration of safety, the thickness of blade part t_w is assumed to be 15 [mm].

[II] Allowable compressive stress :

The allowable compressive stress f_c is assumed to be lower than 1.00×10^6 [Pa].

[III] Maximum torsional torque :

The maximum torsional torque is assumed to be 1.20x10³ [Nm].

[IV] Dimensions of each damper part :

In designing, the dimensions of each damper part have been specified in consideration of compactness and simple structure. The ranges of them are as follows :

[a] Number of rubber block pairs:

$n = 4$ to 12

[b] Shape factor rubber block:

$U = 0.1$ to 0.5

[c] Pitch circular radius of rubber block:

$R = 60$ to 120 [mm]

[d] Thickness of damper inertia ring in radial direction:

$t = 0.6$ to 20 [mm]

[e] Outside diameter of damper inertia:

$d_1 = 150$ to 220 [mm]

[f] Inertia ratio [3]* to [6]*:

$\alpha = 0.1$ to 0.2

The above - mentioned n, U, R, and inertia of damper inertia ring I_d are used as the variable parameter in the design procedure. We have formulated the design calculation program and examines the output result data as mentioned below.

At first, the value of K_d can be obtained by substituting the values of I_d, ω_e, I_e, (ω_e, I_e : known values) into equation (5). The value of I_d is obtained by the design calculation program.

Next, the values of compression spring constant K and shape factor U are given by substituting the set values obtained by the design calculation program into equations (1) and (2), respectively. The obtained values of K and U are compared with the experimental results shown in Figure 3 and are examined.

Finally, the design calculation program examines if the dimensional values of each shown part meet the requirement conditions.

Table 3 Dimensions of New Torsional Vibration Rubber Damper of Compression Type [Calculation Results]

Number	R [mm]	U	a [mm]	b [mm]	l [mm]	d_1 [mm]	d_2 [mm]	I_d [kgm²]	K_d [Nm/rad]
1	70.0	0.30	12.5	40.0	9.0	196.0	180.0	1.098x10⁻²	2.05x10⁴
2	75.0	0.20	14.5	34.0	7.0	200.0	184.0	1.059x10⁻²	1.99x10⁴
3	75.0	0.25	14.5	39.0	9.0	205.0	189.0	1.265x10⁻²	2.28x10⁴
4	75.0	0.30	14.5	29.0	12.0	195.0	179.0	1.098x10⁻²	2.06x10⁴
5	80.0	0.20	16.4	40.0	8.0	215.0	199.0	1.481x10⁻²	2.56x10⁴
6	80.0	0.30	16.4	39.0	13.0	215.0	199.0	1.677x10⁻²	2.79x10⁴

Table 3 shows the dimensional values of damper parts meeting the requirement conditions. Using this table, we have selected the mutually close values of width of rubber block in circumferential direction along pitch line, length of rubber block in radial direction, and length of rubber block in axial direction. Thus, for prototype fabrication, we have adopted dimensional values of **No.4** damper parts in **Table 3** the dimensions of the damper parts except them shown in **Table 3** are as follows:

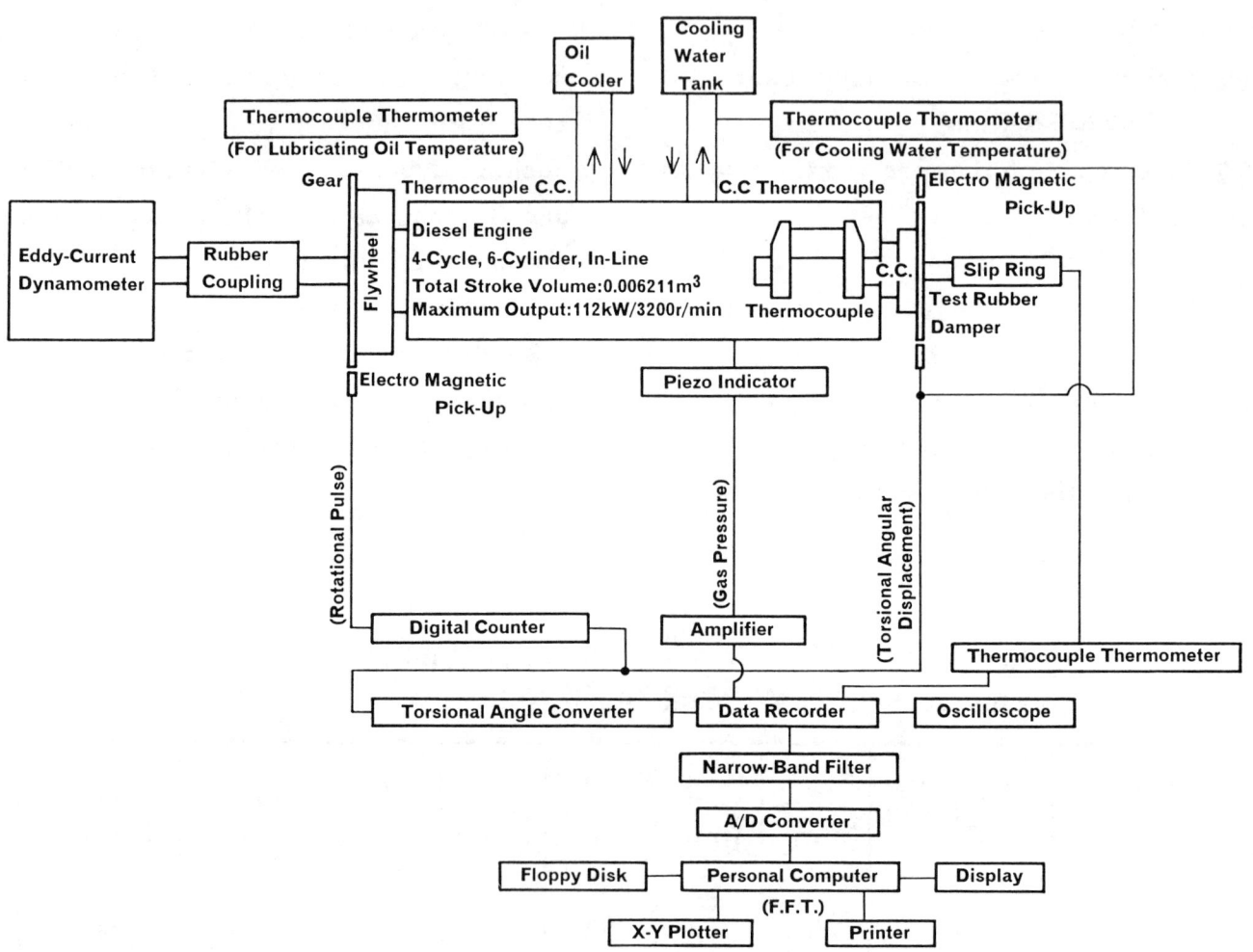

Figure 4: System Diagram of the Engine Test for Crankshaft Torsional Vibration

Number of rubber block pairs :
$n = 8$

Thickness of damper inertia ring in radial direction :
$t = 8.0$ [mm]

Thickness of blade part of damper inertia ring :
$t_w = 15$ [mm]

EXPERIMENT ON ACTUAL ENGINE

As to the test engine crankshafting having the equivalent vibration system shown in Figure 2 and the specifications shown in Table 1, we have designed and fabricated the compression - type rubber damper. Then, we carried out actual engine experiment with it mounted on the crankshaft pulley end.

MEASUREMENT OF TORSIONAL VIBRATION ANGULAR DISPLACEMENT AMPLITUDE - Figure 4 shows the experiment scheme for measurement of torsional vibration angular displacement amplitude.

In this experiment, the eddy - current dynamometer is connected with the test engine via the rubber coupling, and the test compression - type rubber damper is mounted on the pulley end of the crankshafting. The pulse generator gear is attached on each of the housing part and inertia ring part of the compression - type rubber damper, and the frequency signal in proportion to the engine rotational speed is taken out through the electromagnetic pick - up circuit. The output frequency signal is fed to the average angular velocity [center frequency] calculating adapter and then to the phase - shift type torsiograph equipment. Using a phase difference between the detected frequency and the center frequency, a value of torsional vibration angular displacement is calculated to provide a torsional vibration waveform. Furthermore, the torsional vibration waveform data is input to the personal computer for spectral analysis. Table 4 shows the experimental conditions.

Table 4 Experimental Conditions

Items	Values
Engine Speed r/min	1000-3200
Cooling Water Temperature K	353
Lubrication Oil Temperature K	333
Load	1/4 Load

For examination in comparison with the vibration characteristics of the shear - type rubber damper [7]* to [16]*, we have carried out similar experiment on the standard damper actually used in the test engine. During experiment, the cooling water temperature, lubricant temperature and damper rubber temperature have been maintained at constant level.

MEASUREMENT RESULTS OF TORSIONAL VIBRATION ANGULAR DISPLACEMENT AMPLITUDE - Figure 5 shows the amplitude curve of torsional

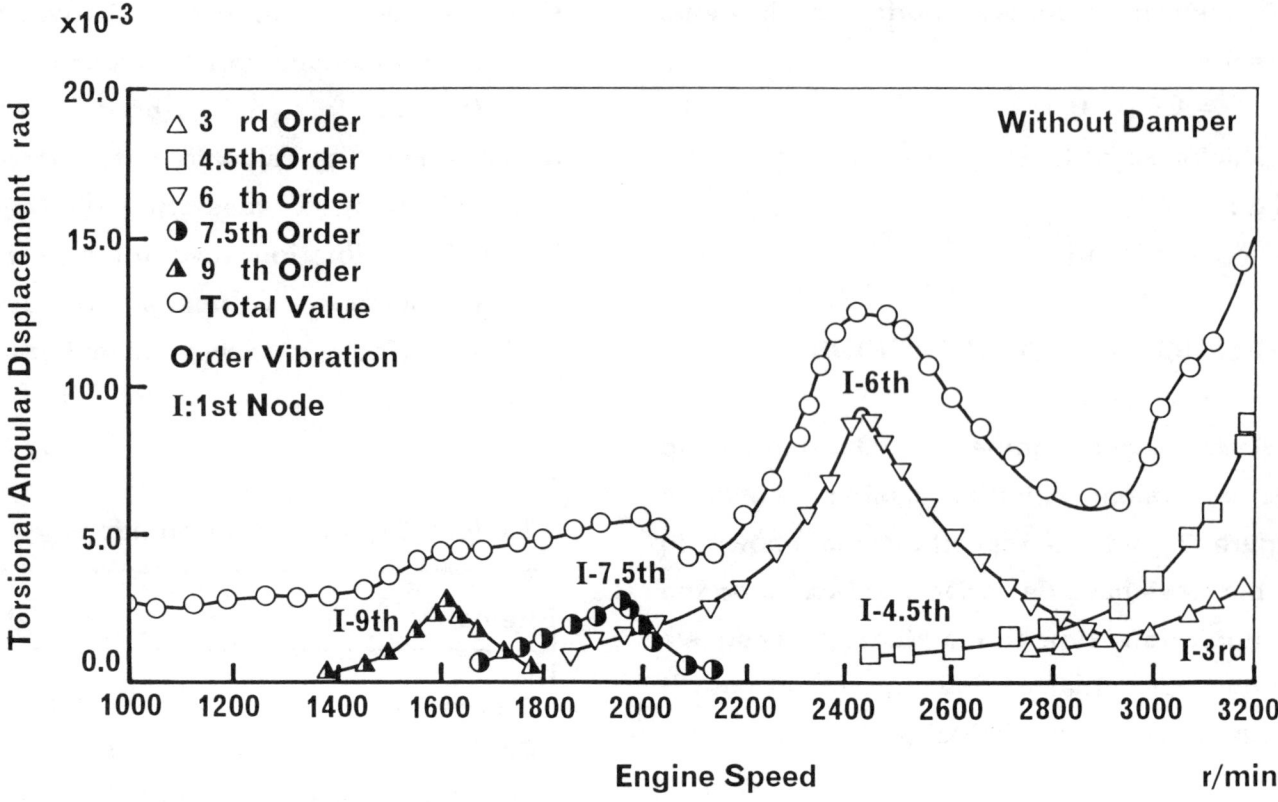

Figure 5: Torsional Vibration Amplitude Curves at Crankshaft Front End of Test Engine without Rubber Damper

vibration angular displacement measured at the end of crankshaft in the engine having no rubber damper. The 6th order vibration resonance point [main critical order component] of the first node vibration is located at 2456 [r/min]. The measured maximum amplitude value is 1.27×10^{-3} [rad], which is appreciably significant with respect to the allowable torsional angular displacement of crankshaft.

Figure 6 shows the amplitude curve of torsional vibration angular displacement measured at the pulley end of the crankshaft equipped with the compression - type rubber damper [NR60], and Figure 7 shows that measured at the pulley end of the crankshaft equipped with the shear - type rubber damper. To provide equivalent conditions for both the dampers in comparison, the additional inertial mass values have been adjusted. Thus, the equivalent tuning ratio has been given to both the dampers.

In examination of the damper characteristics, the 6th order vibration values [main critical order vibration components] of these dampers have been compared. The 6th order resonance point of the first node vibration in the compression - type rubber

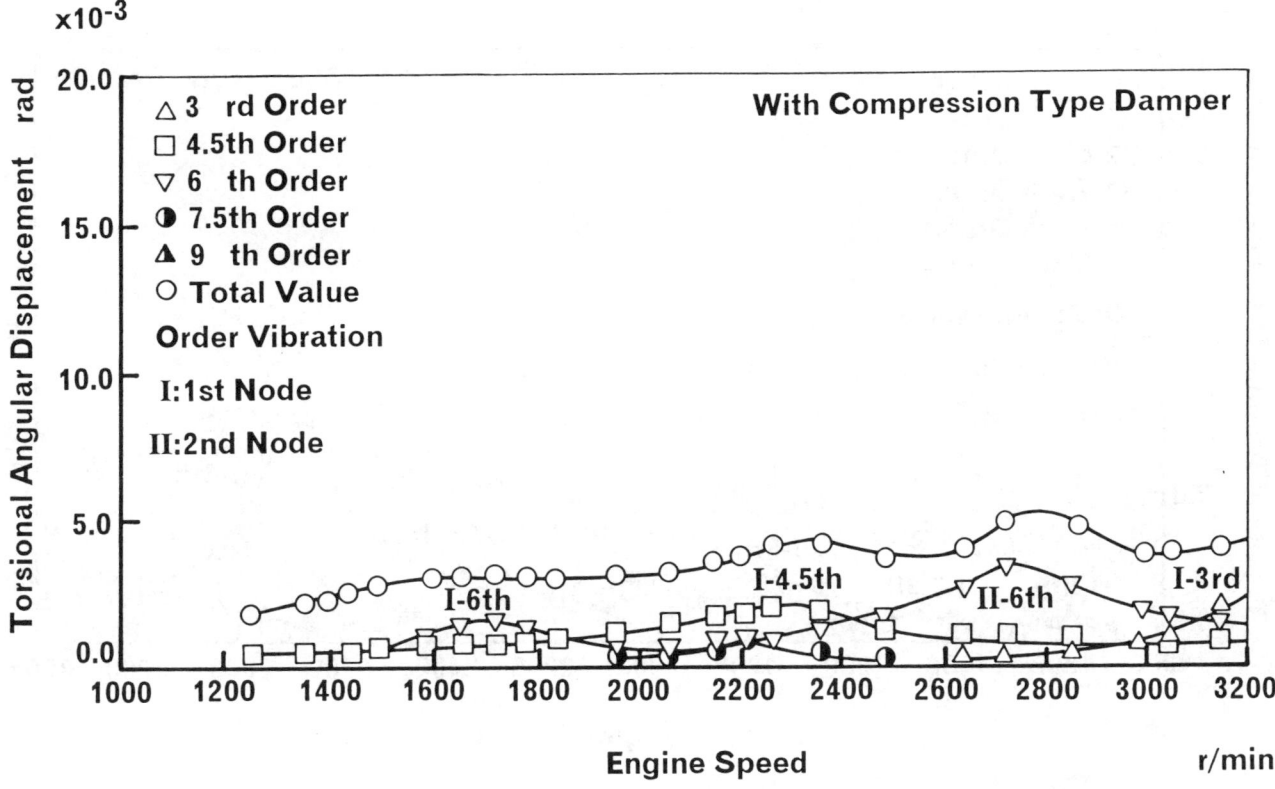

Figure 6: Torsional Vibration Amplitude Curves at Crankshaft Front End with Torsional Vibration Rubber Damper of Compression

damper is located at 1720 [r/min] and the maximum amplitude is 0.32×10^{-3} [rad]. In the shear - type rubber damper, the vibration occurs at 1790 [r/min], and the maximum amplitude is 0.53×10^{-3} [rad]. Also, the 6th order resonant point of the second node vibration in the compression - type rubber damper is located at 2750 [r/min] and the maximum amplitude is 0.57×10^{-3} [rad]. In the shear - type rubber damper, the vibration occurs at 2690 [r/min] and the maximum amplitude is 0.97×10^{-3} [rad].

Thus, in comparison between the compression - type rubber damper and the shear - type rubber damper, it has been found that the reduced effect of vibration amplitude of the former is approximately 40 [%] higher than that of the latter in the operating rotation speed range under the aforementioned conditions.

Also, we have conducted similar experiment on the inertia ring with the measurement gear attached. In the result of this experiment, it has been revealed that the maximum relative amplitude of rubber part [representing an amplitude difference between the housing part and the inertia ring part] in the compression - type rubber

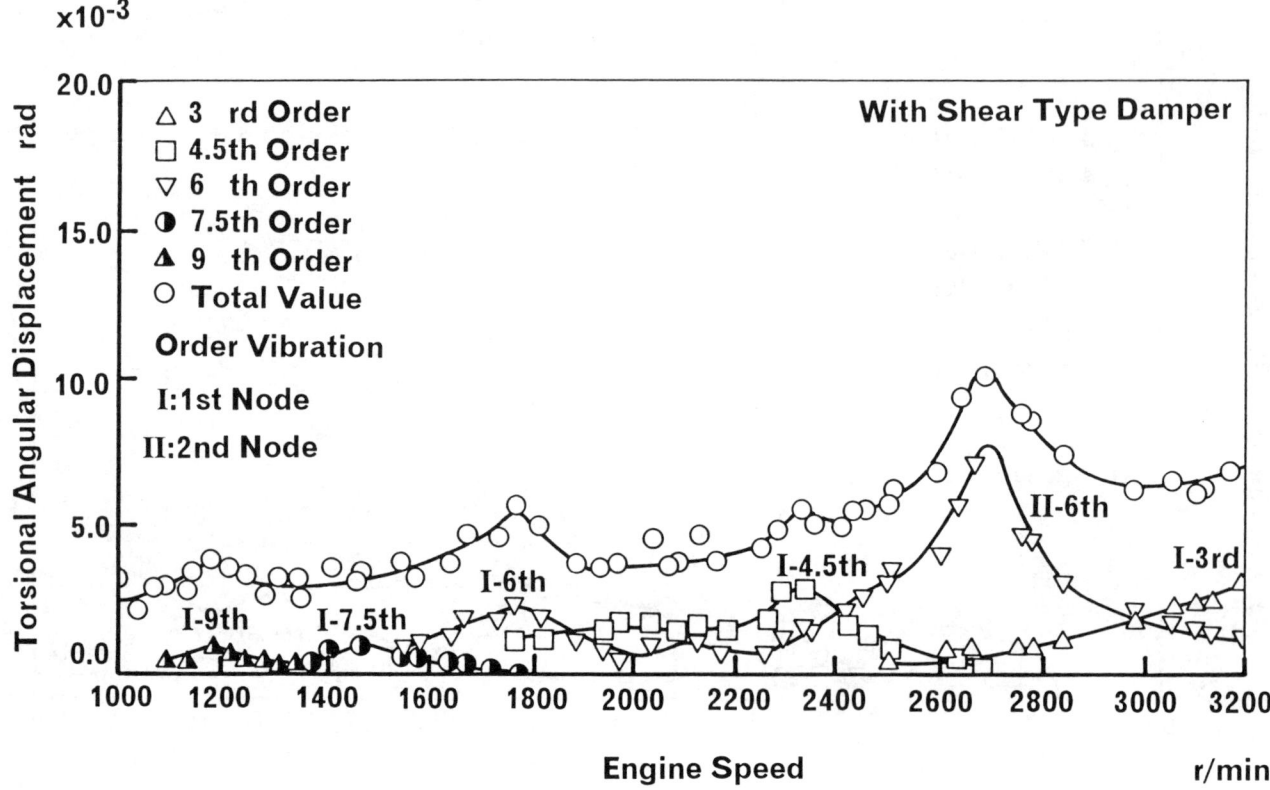

Figure 7: Torsional Vibration Amplitude Curves at Crankshaft Front End with Torsional Vibration Rubber Damper of Shear Type

damper is 0.97×10^{-3} [rad] in the vicinity of 2750 [r/min], and the maximum relative amplitude of rubber part in the shear - type rubber damper is 1.68×10^{-3} [rad] in the vicinity of 2690 [r/min]. Based on these results, we have judged that the strain imposed on the rubber part of the compression - type rubber damper is smaller than that imposed on the rubber part of the shear - type rubber damper. Also, in consideration of the condition that the allowable compressive stress is larger than the allowable shearing stress, the durability of the compression - type rubber damper is thought to be higher than that of the shear - type rubber damper.

EXAMINATIONS

DYNAMIC TORSIONAL STIFFNESS OF RUBBER DAMPER - The engine crankshafting equipped with the rubber damper can be replaced by the two - degree of freedom, equivalent vibration system. In this system, the equation of motion for the inertia ring part is expressed as shown below.

$$I_d \cdot \ddot{\theta}_d + C_d(\dot{\theta}_d - \dot{\theta}_p) + K_d(\theta_d - \theta_p) = 0 \quad (10)$$

Where, θ_p, θ_d and M are defined as follows.

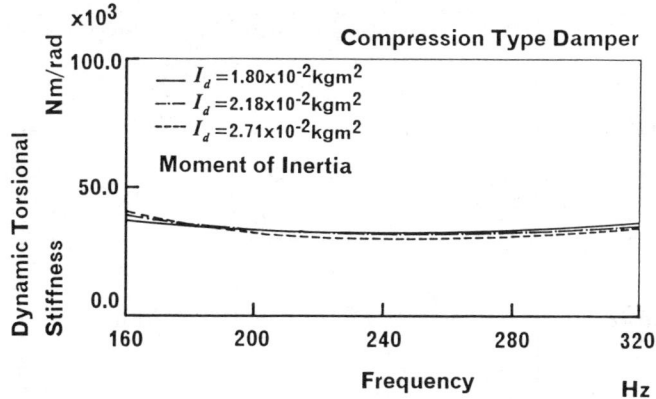

Figure 8: Relationship Between Frequency and Dynamic Torsional Stiffness of Torsional Vibration Rubber Damper of Compression Type

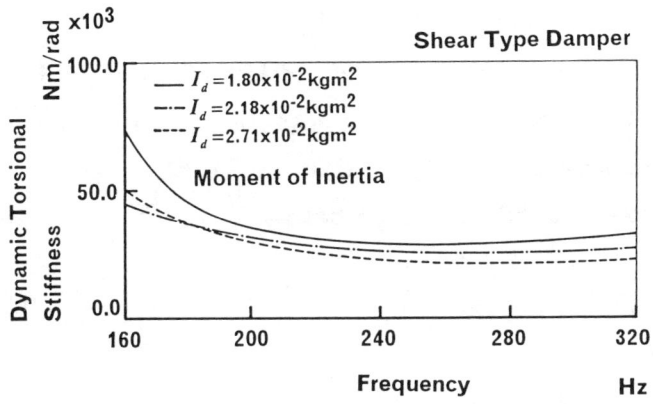

Figure 9: Relationship Between Frequency and Dynamic Torsional Stiffness of Torsional Vibration Rubber Damper of Shear Type

$$\theta_p = \theta_{po} e^{j\omega t} \quad (11)$$

$$\theta_d = \theta_{do} e^{j(\omega t - \phi)} \quad (12)$$

$$M = \frac{\theta_{do}}{\theta_{po}} \quad (13)$$

By arranging equation (10) into the real and imaginary parts using equations (11) to (13), the torsional stiffness and the damping coefficient, which are representing the dynamic characteristic of damper, can be attained as shown below.

$$K_d = \frac{\omega^2 \cdot I_d \cdot M(M - \cos\phi)}{M^2 + 1 - 2 \cdot M \cdot \cos\phi} \quad (14)$$

$$C_d = \frac{\omega \cdot I_d \cdot M \cdot \sin\phi}{M^2 + 1 - 2 \cdot M \cdot \cos\phi} \quad (15)$$

Figure 8 shows the relationship between the dynamic torsional stiffness and frequency of the compression - type rubber damper attained by substituting the experimental result data in equation (14) and Figure 9 shows that of the shear - type rubber damper. As the inertia moment of the damper inertia ring is changed in the experiment, the vibration system is changed. Under this condition, the effect of the frequency on the torsional stiffness of rubber damper has been examined.

In examination, it has been found that the dependency of the frequency on the dynamic torsional stiffness in the compression - type rubber damper is smaller than that in the shear - type rubber damper. This characteristic of the compression - type rubber damper is advantageous in practical designing.

Table 5 Dynamic Torsional Stiffness and Dynamic Magnifier of Stiffness

Damper Name	1st Node, 6th Order Vibration		2nd Node, 6th Order Vibration	
	Dynamic Torsional Stiffness K_d Nm/rad	Dynamic Magnifier of Stiffness	Dynamic Torsional Stiffness K_d Nm/rad	Dynamic Magnifier of Stiffness
Shear Type Rubber Damper [NR60]	3.070×10^4	2.302	2.262×10^4	1.696
Compression Type Rubber Damper [NR60]	2.767×10^4	1.484	2.494×10^4	1.377

NR60:Natural Rubber, Shore Hardness:Hs=60

DYNAMIC MAGNIFIER OF STIFFNESS AND DYNAMIC TORSIONAL STIFFNESS - Table 5 shows the dynamic torsional stiffness values and dynamic magnifier of stiffness [ratios of dynamic torsional stiffness to static torsional stiffness] that have been calculated for the 6th order vibration components of the 1st and 2nd node using the inverse Holtzer method. A change in the dynamic torsional stiffness of the compression - type rubber damper is smaller than that in the shear - type rubber damper, and the dynamic magnifier of stiffness in the former is also smaller than that in the latter. This means that the compression - type rubber damper is advantageous for torsional stiffness setting in damper designing.

LOSS ENERGY OF RUBBER DAMPER - The loss energy representing the energy absorbed by the damper rubber part is expressed as shown below.

$$H = \pi \cdot K_d \cdot \omega \cdot \tau \cdot \theta_{RO}^2 \qquad (16)$$

In the above equation, τ indicates C_d / K_d. Since the loss energy absorbed by the rubber part is converted into thermal energy, it has an effect on the durability of the rubber part. Figures 10 and 11 show the loss energy curves of the compression - type and shear - type rubber dampers with respect to the frequency.

In comparison of the loss energy values in the vicinity of the 6th order resonance point of the 2nd node, it has been found that the

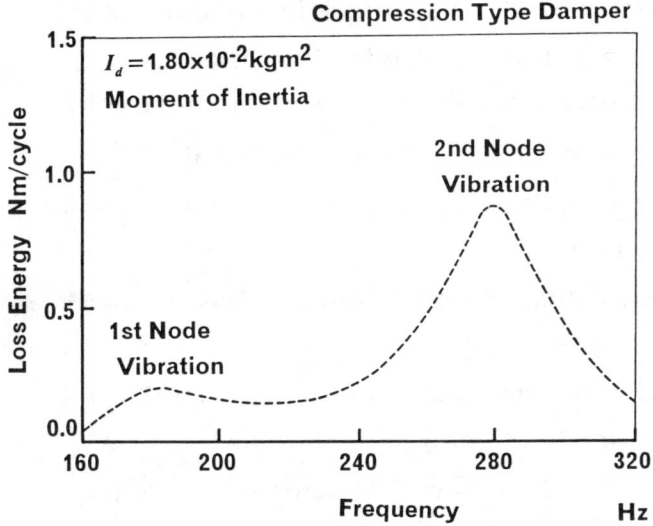

Figure 10: Loss Energy Curve of Torsional Vibration Rubber Damper of Compression Type

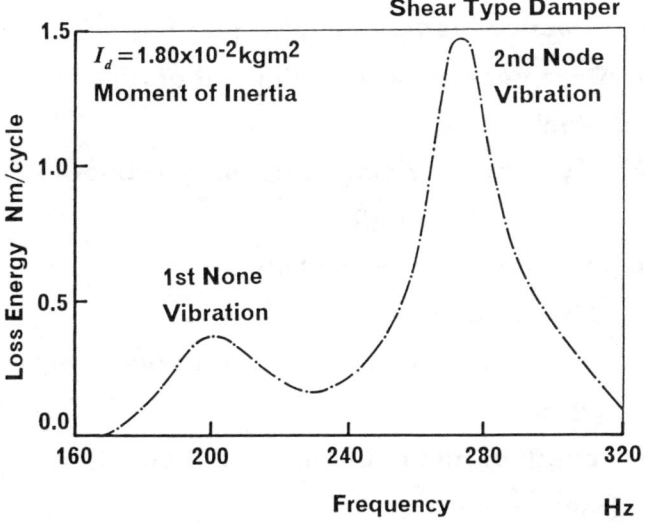

Figure 11: Loss Energy Curve of Torsional Vibration Rubber Damper of Shear Type

loss energy of the compression - type rubber damper is approx. 40 [%] smaller than that of the shear - type rubber damper. This contributes to reduction of thermal load exerted on the rubber part of the compression - type damper. Having this characteristic, the compression - type rubber damper provides superior durability to the shear - type rubber damper.

CONCLUSIONS

As mentioned so far, we have carried out experiments of the newly designed compression - type torsional vibration rubber damper and the conventional shear - type torsional vibration rubber damper using the actual automobile high - speed diesel engine. Summarized below are the results of our research and examination of the vibration characteristics of these rubber dampers.

[1] For reducing the drawback of the conventional shear - type torsional vibration rubber damper, we have formulated the design method for the new compression - type torsional vibration rubber damper.

[2] Under condition of the equal damper tuning ratio, we have measured the torsional vibration angular displacement amplitude of each of the compression - type and shear - type rubber dampers. In comparison of these dampers, it has been revealed that the reduced effect of torsional vibration amplitude of the compression - type rubber damper is higher than that of the shear - type rubber damper. Also, in comparison of relative amplitude levels of the damper rubber parts, it has been found that the relative amplitude value of the compression -

type rubber damper is smaller than that of the shear - type rubber damper. Therefore, the durability of the former damper is assumed to be higher than that of the latter damper.

[3] In vibration characteristic comparison between the compression - type and shear - type rubber dampers, the dependability on the frequency in the compression - type rubber damper has been found to be smaller than that in the shear - type rubber damper. Since the dynamic magnifier of stiffness of the compression - type rubber damper is also smaller, it is advantageous for practical designing.

[4] The compression - type rubber damper provides smaller loss energy than the shear - type rubber damper. Therefore, the durability of the compression - type rubber damper is superior to that of the shear - type rubber damper, i.e. the compression - type rubber damper is advantageous for improvement of durability.

Although the compression - type rubber damper has the above excellent characteristics, it is necessary to solve such matters concerning structural complexity, permanent distortion of rubber block in press - fitting state, manufacturing cost, etc. For solution of these matters, we will continue our research intensively further.

DEFINITIONS OF SYMBOLS

Defined below are the symbols used in this research report.

a : Width of rubber block in circumferential direction along pitch line [mm]
a_1 : Width of rubber block in circumferential direction in natural state [mm]
b : Length of rubber block in radial direction [mm]
C_d : Damping coefficient of rubber damper [Nms/rad]
d_1 : Outside diameter of damper inertia ring ($=2r_1$) [mm]
d_2 : Inside diameter of damper inertia ring ($=2r_2$) [mm]
E_c : Compressive elastic modulus of rubber block [N/m^2]
f_c : Allowable compressive stress [N/m^2]
H : Loss energy [Nm/cycle]
H_s : Shore hardness of rubber
I_d : Inertia of damper inertia ring [kgm^2]
I_e : Effective inertia of engine part [kgm^2]
K : Compression spring constant of rubber block [N/m]
K_d : Dynamic torsional stiffness of rubber damper [Nm/rad]
K_e : Torsional stiffness of engine part [Nm/rad]
K_s : Static torsional stiffness of rubber damper [Nm/rad]
l : Length of rubber block in axial direction [mm]
M : Amplitude ratio ($=\theta_d/\theta_p$)
n : Number of assembled rubber block pairs
R : Radius of pitch circle of rubber block [mm]
r_1 : Outside radius of damper inertia ring [mm]
r_2 : Inside radius of damper inertia ring [mm]
t : Thickness of damper inertia ring in radial direction ($=r_1-r_2$) [mm]
t_w : Thickness of blade part of damper inertia

ring [mm]
U :Shape factor of rubber block
α :Inertial ratio $(=I_d / I_e)$
θ_d :Torsional vibration angular displacement amplitude of damper inertia ring [rad]
θ_p :Torsional vibration angular displacement amplitude of damper housing [rad]
θ_R :Relative torsional angular displacement amplitude between damper inertia ring and damper housing $(\theta_d - \theta_p)$ [rad]
λ :Tuning ratio $(=\omega_d / \omega_e)$
λ_{opt} :Optimum tuning ratio
τ :Ratio of damping coefficient to torsional stiffness $(=C_d / K_d)$ [s]
ϕ :Phase difference between damper inertia ring part and damper housing part [rad]
ω :Angular velocity [rad/s]
ω_d :Natural angular velocity of damper part [rad/s]
ω_e :Natural angular velocity of engine part [rad/s]

ACKNOWLEDGMENTS

The authors would like to express grateful acknowledgment to Mr.Hidenobu Aoki and Mr.Hisashi Hagiwara of Fuji Automobile Ind. Co. for collaboration in fabrication of the compression - type rubber damper.

REFERENCES

[1] Ker Wilson, <u>Practical Solution of Torsional Vibration Problem</u>, Vol.I, Chapman & Hall, 1958.

[2] Ker Wilson, <u>Practical Solution of Torsional Vibration Problem</u>, Vol.II, Chapman & Hall, 1963.

[3] Y. Honda, *et al.*, " Design of New Rubber Damper for Automobile Diesel Engine (in Japanese), " *Proceedings of the 240th Kansai Conference of Japan Society of Mechanical Engineers*, No.203, Japan Society of Mechanical Engineers, Tokyo, Japan, 1979.

[4] T.Saito・Y.Honda, " Characteristics of a New Compression Type Rubber Damper for Controlling Torsional Vibration in a High Speed Multi Cylinder Diesel Engine (in Japanese), " *Bulletin of Science and Engineering Research Laboratory, Waseda University*, No.95, P34-40, Tokyo, Japan, 1989.

[5] T.Seki・T.Saito・S.Iwamoto・N.Eguchi・K.Wakabayashi, " Characteristics of a Rubber Torsional Vibration Damper Attached to a High Speed Multi Cylinder Diesel Engine (in Japanese), " *Bulletin of Science and Engineering Research Laboratory, Waseda University*, No.38, P28, Tokyo, Japan, 1983.

[6] B.I.C.E.R.A., " <u>A Handbook on Torsional Vibration</u>, Cambridge at the University Press, London, U.K., 1958.

[7] S.Iwamoto, K.Wakabayashi, T. Kodama, " Dynamic Characteristics of Torsional Rubber Dampers in High Speed Diesel Engine, " *The Scince and Engineering Reports of Saitama University*, Series C, No.17, P1-7, Saitama, Japan, 1983.

[8] Y.Honda, T.Saito, K.Wakabayashi, T.Kodama, S.Iwamoto, "A Simulation Method for Crankshaft Torsional Vibration by Considering Dynamic Characteristics of Rubber Dampers, " *Proceedings of the 1989 Noise and Vibration Conference*, P-222, P453-469, SAE Paper No.891172, Society of Automotive Engineers, Inc., MI, U.S.A., 1989.

[9] Y.Honda, T.Saito, K.Wakabayashi,

T.Kodama, " A Study on a Simulation Method for Torsional Vibration of Diesel Engine Crankshaft with a Rubber Damper (in Japanese), " *Bulletin of Science and Engineering Research Laboratory, Waseda University*, No.128, P43-57, Tokyo, Japan, 1990.

[10] Y.Honda, T.Saito, K.Wakabayashi, T.Kodama et al., " An Analysis on Torsional Vibration of Crankshaft with Rubber Damper by Using Transition Matrix Method - 1st Report : An Evaluation on Torsional Vibration Rubber Damper with Various Characteristics - (in Japanese), " *Transactions of the Society of Automotive Engineers of Japan*, No.46, P79-84, Society of Automotive Engineers of Japan, Tokyo, Japan, 1990.

[11] T.Kodama, K.Wakabayashi, Y.Honda et al., " An Analysis on Torsional Vibration of Crankshaft with Rubber Damper by Using Transition Matrix Method - 2nd Report : Influence of Temperature Dependency of Rubber Damper on Torsional Vibration (in Japanese), " *Transactions of the Society of Automotive Engineers of Japan*, No.46, P85-89, Society of Automotive Engineers of Japan, Tokyo, Japan, 1990.

[12] K.Wakabayashi, Y.Honda, T.Kodama et al., " Torsional Vibration Damping of Diesel Engines with a Rubber Damper Pulley (in Japanese), " *Transactions of the Japan Society of Mechanical Engineers*, Vol.60, No.572, P1167-1174, Japan Society of Mechanical Engineers, Tokyo, Japan, 1994.

[13] T.Kodama, K.Shimoyamada, S.Iwamoto, Y.Honda, K.Wakabayashi et al, " A Numerical Analysis of Forced Torsional Vibration of Crankshaft with a Shear Type Rubber Torsional Damper by Transfer Matrix Method - 1 st Report : Comparison with Computation and Measurement of Torsional Vibration Displacement - (in Japanese), " *Transactions of the Society of Automotive Engineers of Japan*, Vol.23, No.1, P106-112, Society of Automotive Engineers of Japan, Tokyo, Japan, 1992.

[14] K.Shimoyamada, T.Kodama, Y.Honda, K.Wakabayashi, S.Iwamoto, " A Numerical Computation for Vibration Displacements and Stresses of a Crankshaft with a Shear Rubber Torsional Damper, " SAE Paper No.930197, P1-21, MI, U.S.A. 1993.

[15] K.Wakabayashi et al. , " Torsional Vibration Analysis of Crankshaft with Rubber Damper (in Japanese), " *Journal of the Society of Automotive Engineers of Japan*, Vol.35, No.12, P1423-1427, Society of Automotive Engineers of Japan, Tokyo, Japan, 1981.

[16] K.Wakabayashi et al., " A Simulation Method for Torsional Vibration Waveform of a Crankshaft with a Rubber Damper, " *I.Mech.E. Conference Publication*, C119 / 79, P21-31, I.Mech.E. London, U.K., 1979.

950524
Novel Approach to Reduce the Time from Concept-to-Finished Piston

Jeffrey L. Castleman and Darren R. Bailey
Zollner Corp.

ABSTRACT

An increasing emphasis is currently being placed on reducing the design-to-manufacturing time for all products. Pistons for internal combustion engines are being placed under the same emphasis. A new method has been established which reduces the piston design-to-manufacturing time and thus, brings pistons to market quicker than ever before.

This new method involves performing the four standard steps to any product design-to-manufacturing sequence. The first step is to design the product from an initial concept. The second step is to design the manufacturing tooling necessary to produce this product. The third step is to build this manufacturing tooling and the final step is to produce the product using this tooling. The new method now involves the use of a 3-D solid modeling CAD/CAM system to design the pistons and the manufacturing tooling necessary to make these pistons. This piston information is used to produce 3-D solid models of the tooling necessary to manufacture this particular piston. The tooling geometry information is then passed to the tooling manufacturers who use this information to build the piston tooling directly through the use of CNC machines. To help shorten the tool build, all standard tooling components have been pre-machined so that the only critical machining added is the piston geometry. Once the tooling is complete, the pistons are manufactured in a flexible prototype environment which utilizes standard tooling and procedures.

The effective use of 3-D solid modeling CAD/CAM systems and the use of standard tooling has reduced the time necessary to perform each of the design-to-manufacturing steps for a new piston concept. By reducing the time necessary to perform each of these steps, the overall concept-to-finished piston time has been reduced by 60%.

INTRODUCTION

When establishing additional business with new or existing customers, design-to-manufacturing time is important for any manufacturing organization. In today's market, the first organization to produce the product will most likely receive the new business. For piston manufacturers, designing a new piston and its respective mold and manufacturing this piston mold are the more crucial steps in the production of prototype pistons for customer needs. Figure 1 illustrates the piston prototyping flow chart which was used for many years previously to achieve a successful piston design-to-manufacturing process. It involved the use of two-dimensional drawings of the piston and its mold and the skillful "artistic" interpretation of these blueprints to manufacture the mold equipment. The entire process typically took 21 weeks from start to completion. Once a new computer aided design/computer aided manufacturing system (CAD/CAM) was implemented, the piston design-to-manufacturing time decreased substantially. Figure 2 exhibits the new piston rapid prototyping flow chart. This new procedure centers around the use of three-dimensional solid models of the piston and its mold along with the use of CAM toolpathing and CNC milling technology to manufacture the piston mold on pre-built mold components. This new procedure has significantly reduced the time from concept-to-finished piston from 21 weeks to 12 weeks. The following sections of this report provide more information on each of the major steps in this novel piston rapid prototyping procedure.

CUSTOMER REQUIREMENTS

The main objective of a new piston design is that it meets all the design goals established by the customer. These goals include cost effective, lightweight, emission sensitive, durable, low friction, low wear, low noise, etc. Figure 3 illustrates how each of these goals effect the four major engine categories. These goals are ordered with the most important on top of the list and the least important on the bottom.

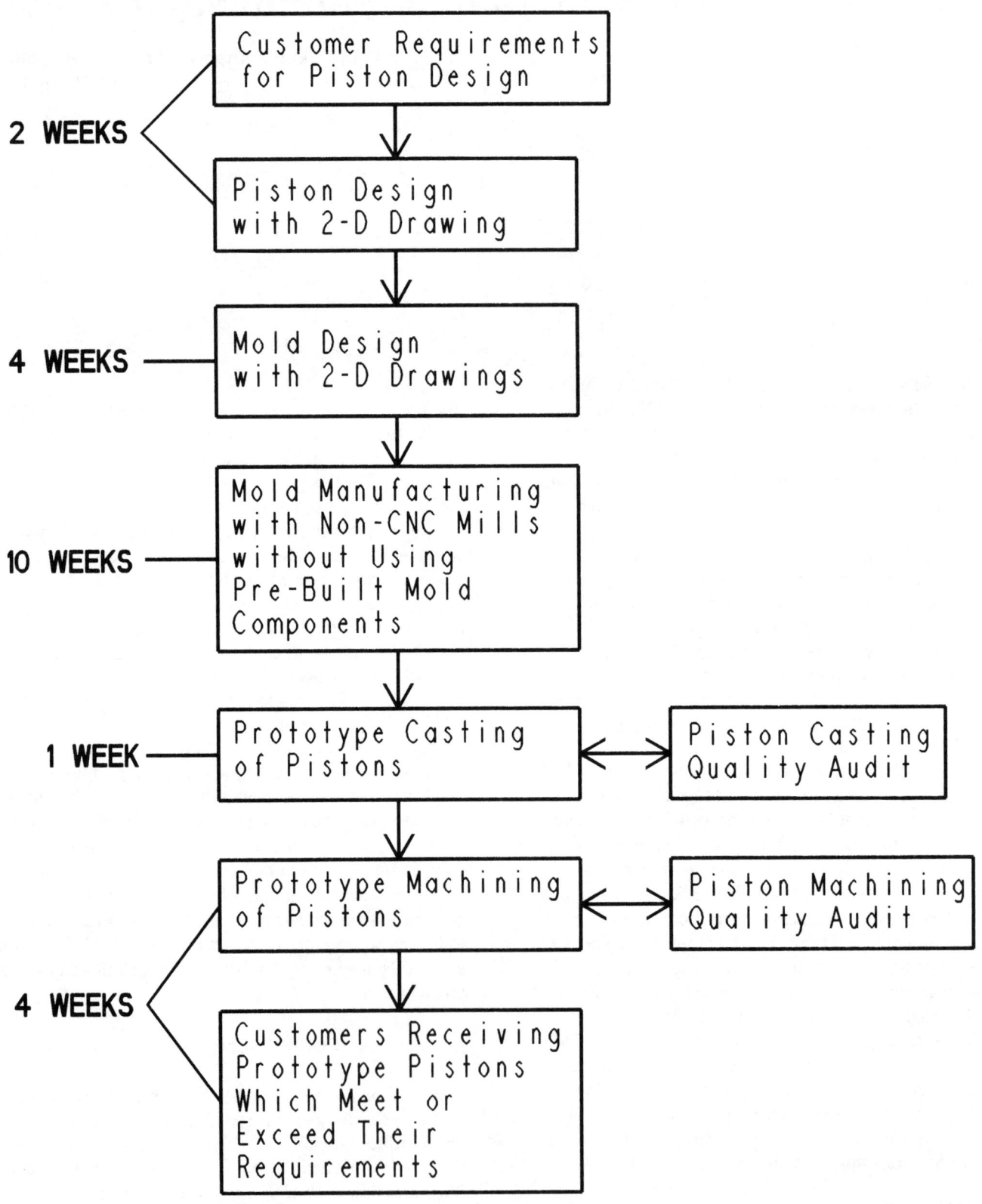

Figure 1

New Piston Rapid Prototyping Flow Chart

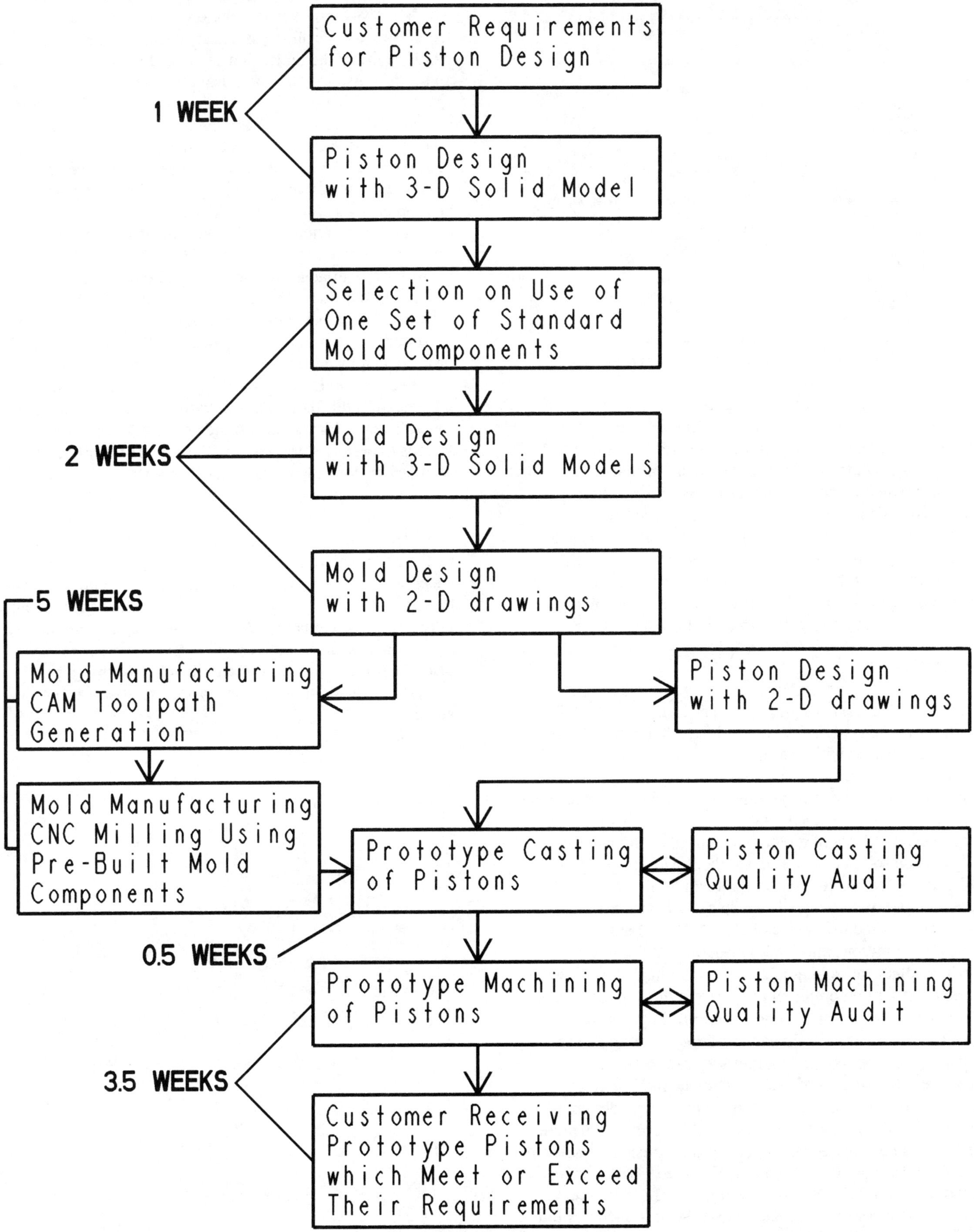

```
                    ┌─────────────────────────┐
                    │ Customer Requirements   │
                    │ for Piston Design       │
  1 WEEK            └───────────┬─────────────┘
                                ▼
                    ┌─────────────────────────┐
                    │ Piston Design           │
                    │ with 3-D Solid Model    │
                    └───────────┬─────────────┘
                                ▼
                    ┌─────────────────────────┐
                    │ Selection on Use of     │
                    │ One Set of Standard     │
                    │ Mold Components         │
                    └───────────┬─────────────┘
                                ▼
                    ┌─────────────────────────┐
  2 WEEKS           │ Mold Design             │
                    │ with 3-D Solid Models   │
                    └───────────┬─────────────┘
                                ▼
                    ┌─────────────────────────┐
                    │ Mold Design             │
                    │ with 2-D drawings       │
                    └───┬─────────────────┬───┘
                        │                 │
  5 WEEKS               ▼                 ▼
        ┌────────────────────┐    ┌────────────────────┐
        │ Mold Manufacturing │    │ Piston Design      │
        │ CAM Toolpath       │◄───│ with 2-D drawings  │
        │ Generation         │    └──────────┬─────────┘
        └─────────┬──────────┘               │
                  ▼                          │
        ┌────────────────────┐               │
        │ Mold Manufacturing │               │
        │ CNC Milling Using  │               │
        │ Pre-Built Mold     │               │
        │ Components         │               │
        └─────────┬──────────┘               │
                  ▼                          ▼
              ┌──────────────────┐    ┌──────────────────┐
              │ Prototype Casting│◄──►│ Piston Casting   │
              │ of Pistons       │    │ Quality Audit    │
  0.5 WEEKS   └────────┬─────────┘    └──────────────────┘
                       ▼
              ┌──────────────────┐    ┌──────────────────┐
              │ Prototype Machin-│◄──►│ Piston Machining │
              │ ing of Pistons   │    │ Quality Audit    │
              └────────┬─────────┘    └──────────────────┘
                       ▼
              ┌──────────────────────────┐
  3.5 WEEKS   │ Customer Receiving       │
              │ Prototype Pistons        │
              │ which Meet or Exceed     │
              │ Their Requirements       │
              └──────────────────────────┘
```

Figure 2

```
Customer Requirements
   Piston Design

Diesel Engine - Off-Highway        Diesel Engine - On-Highway

Durability                         Durability
Fuel Economy                       Emissions
Emissions                          Noise
Noise                              Fuel Economy
Performance                        Performance
Cost                               Cost
Weight                             Weight

Automotive Engine                  Small 4-Stroke  & 2-Stroke

Cost                               Cost
Emissions                          Performance
Performance                        Durability
Noise                              Noise
Fuel Economy                       Emissions
Weight                             Fuel Economy
Durability                         Weight
```

Figure 3 - Customer Requirements for Piston Design

Most piston designs starts with some main dimensions that need to be satisfied. These include bore diameter, compression height, pin diameter and length and ring groove geometry. Connecting rod and crankshaft geometry can be used to guide the design of the piston under-crown and skirt open end geometry, respectively. The engine manufacturer needs to specify the piston top geometry such as combustion bowl, dishes and/or domes. All this information must be specified at the concept stage so that the piston design can be created. If any information is missing, the overall rapid prototyping time will be extended.

PISTON DESIGN

The piston design process has changed drastically with the use of computer-aided design technology. The age of designing pistons with two-dimensional drawing board and CAD drafting drawings has passed and a new three-dimensional piston solid model design has arrived. The piston solid model is the foundation upon which all further processes in this rapid prototyping procedure are based. This piston solid model has come about with the staggering advances in both computer hardware and software. Engineering workstations have replaced mainframe computers as the workhorse of the CAD/CAM systems. These workstations allow solid modeling CAD software to operate effectively and quickly. The solid modeling software used for this rapid prototyping process is Pro/Engineer from Parametric Technology Corporation. Pro/Engineer has several features that facilitate the 3-D piston design. These include solid modeling, feature-based, parametric, full associativity and ease-of-use. Solid modeling is a process of using a primitive volume, such as a cube or cylinder, as a starting shape of a particular design. This primitive volume is then changed by employing cuts and protrusions to achieve the final configuration of the design desired. The solid model produces a solid object with its entire outer boundary fully defined with accurate surfaces. Since the solid model is fully defined, accurate weight and engineering properties of the design can easily be calculated. Feature-based is a method of ordering the solid modeling process. Each individual cut and protrusion of the design process has its own identification which allows for easy manipulation, such as deletion, reordering, etc. Feature based also allows for easy modification of the dimension associated with a particular cut and/or protrusion. This modification process is termed "parametric" due to the fact that each individual dimension of a feature has its own unique name with a value assigned to that name. Modification then becomes merely a procedure of assigning a different value to that parameter (dimension name). Full associativity is a process of allowing geometry and dimensions of a design to be used in all aspects of the design representation. This implies that geometry and dimensions associated with the solid model are also the same geometry and dimensions found in the 2-D drawing of the design. Combined with the parametric nature of this software, full associativity provides that a change made to the solid model will also be automatically reflected in the 2-D drawing and vice versa. Ease-of-use comes from the simple but effective nature of the graphical user interface employed for operating this particular software. The above features in this software package allow design engineers at Zollner Pistons to design every piston with the total definition and accuracy of three dimensional CAD models.

The solid modeling approach to piston design starts with a cylinder of the overall piston casting length and diameter. The piston cylinder is then hollowed to form the piston under-crown and skirt thickness. Protrusions are added for the ring belt, skirt open end belts and pin boss with its support(s). Cuts are made for the skirt tails, pin boss outer reliefs and cast pin bores. All cast piston top features are added along with the radii on all the sharp corners of the piston casting. At this moment, the piston casting is complete. Illustrations of a cast piston are given in figures 4 and 5.

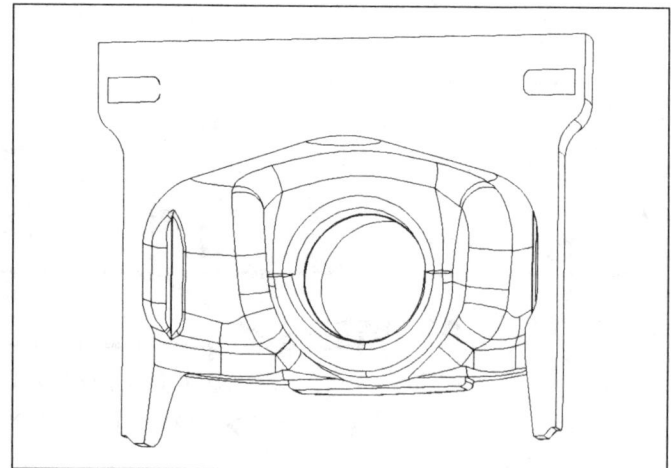

Figure 4 - Interior State of Cast Piston

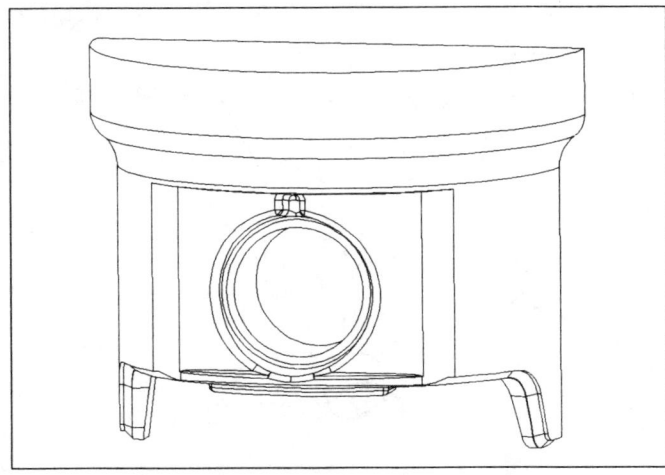

Figure 5 - Exterior State of Cast Piston

These cast features will be used during the piston mold design discussed later. The piston casting is then subjected to all the machining cuts necessary to produce a finished piston. These include ring grooves, skirt and lands, piston top features, such as bowls or dishes, and pin bores as shown in figures 6 and 7.

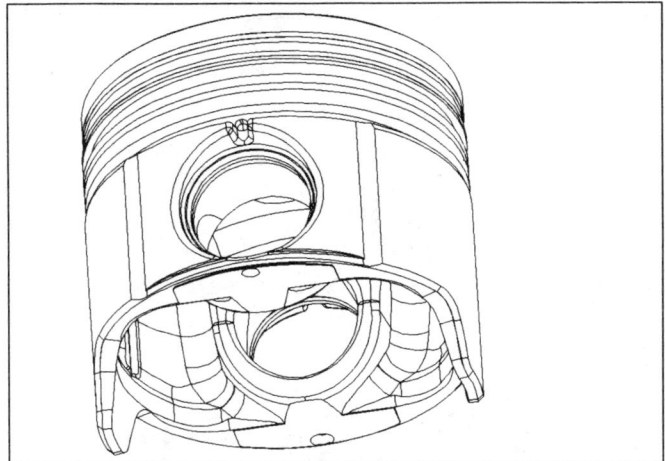

Figure 6 - Finished Piston-View #1

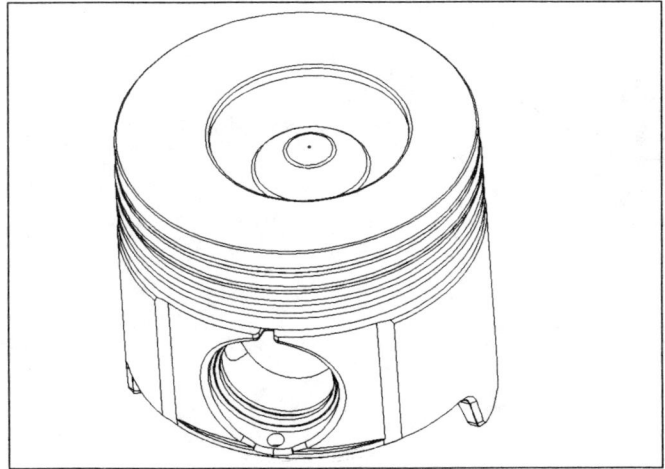

Figure 7 - Finished Piston-View #2

Once complete, the "finished" piston has now been successfully created as a computer solid model. The dimensions used to create the solid model are determined from the customer requirements phase of this process as well as design standards developed through years of piston designing. The solid model is then used to create the views and dimensions necessary for an accurate 2-D drawing of the piston. The piston drawing is completed after the mold design process has taken place, which will be described in detail next. The 2-D drawing is only used as a reference for inspecting and manufacturing the pistons in the prototype production phase of this piston rapid prototyping procedure.

MOLD DESIGN

Once a piston has been designed, a mold must be designed and manufactured in order to produce the piston. With the use of the new CAD/CAM system, the mold design now uses the piston 3-D solid model to transfer geometrical information to the various mold components. Figure 8 illustrates the many components of a mold and the names of each component.

With the new CAD/CAM system, mold design begins with the selection of the proper pre-designed mold components. These components are stored in a computer library, which consists of a variety of pre-mold designs to accommodate different types of pistons. Since the design of this piston requires a ring insert, the mold must be designed with the use of a pivot cap to allow for the insertion of the ring insert. Included with the pivot cap design will be a center riser. A center riser design will have a simple gate, as shown in figure 9, on the side of the piston with the riser centered on top of the piston.

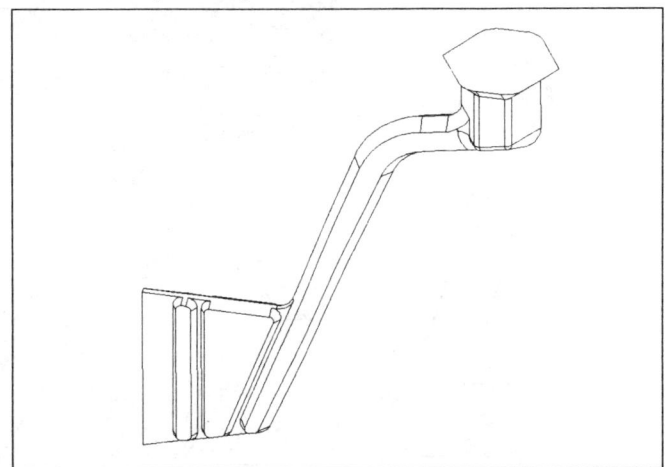

Figure 9 - Gate Design

By having a computer library of pre-designed molds, a designer can take all pre-designed components of the mold and modify them for their piston. Because of the parametric capability of Pro/Engineer, it is possible to modify specific features to make the mold halves fit each new piston design. The mold halves consist of the following parts: bottom split, top split, gate block, mold boss insert (MBI) and core pin. Figures 10 illustrates the bottom split, top split, gate block and mold boss insert components of a mold half.

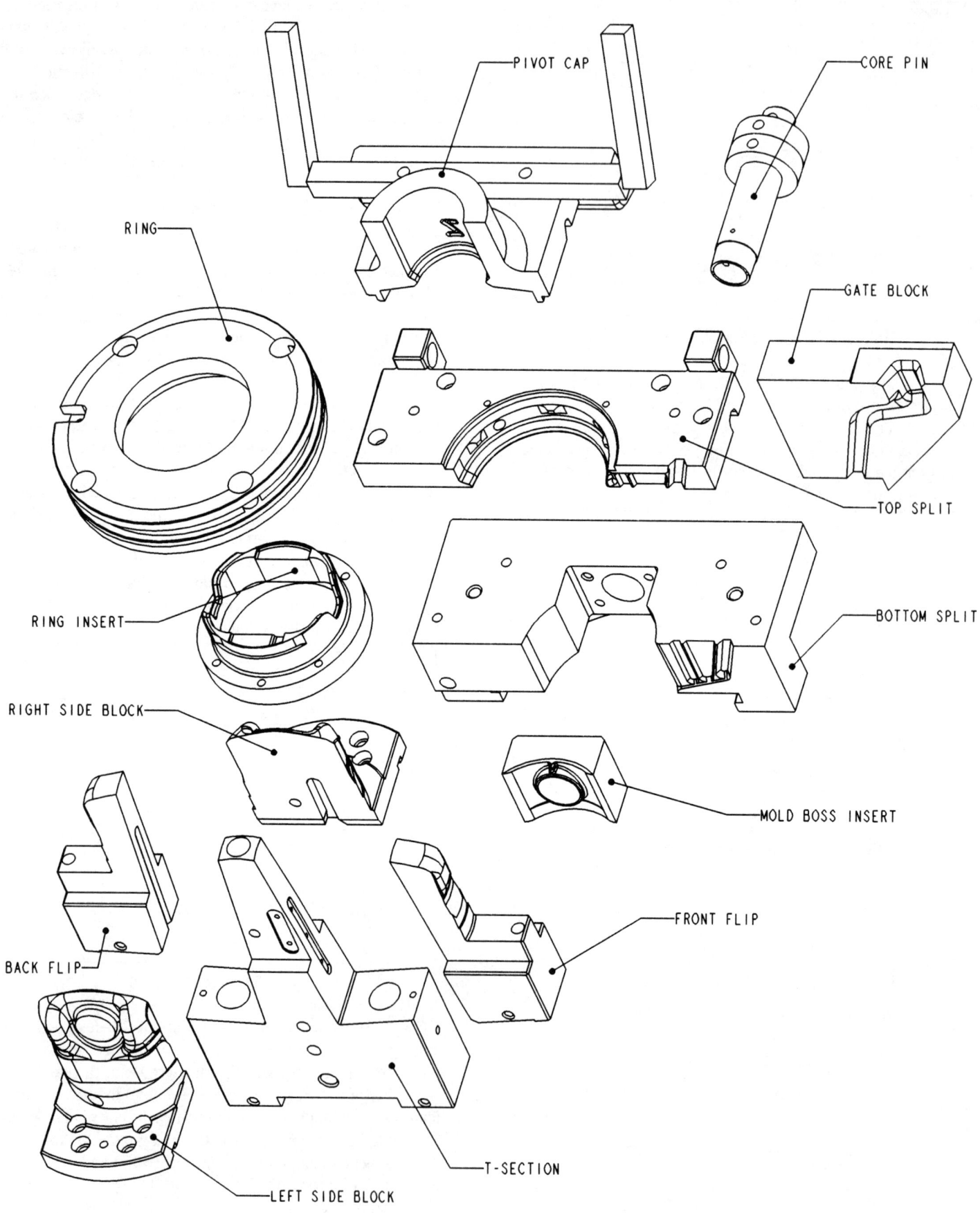

Figure 8 - Exploded Detail of Entire Mold

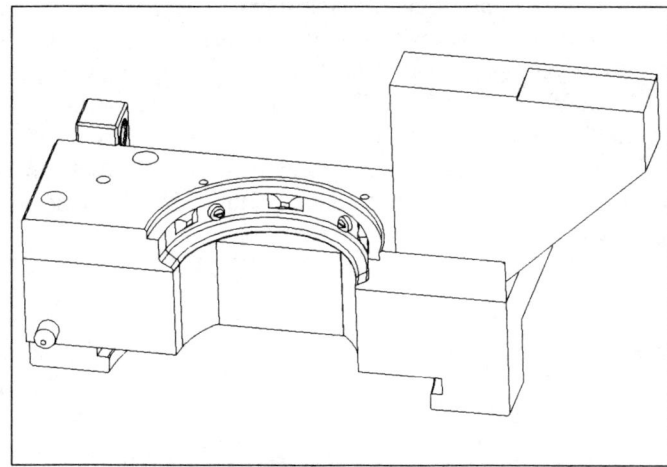

Figure 10 - Bottom Split, Top Split, MBI, Gate Block

The bottom split is the major component of the mold half for it forms a major portion of the exterior of the piston. The MBI and core pin will also be located within the bottom split. The bottom split, when the mold is closing, is oriented by sliding over the ring assembly. Cooling must be incorporated into the mold to speed the solidification of the metal. In order to accomplish this, water lines are placed in the mold. The top split is added to the mold halves so that any necessary cooling lines can be located between the top and bottom split. With the necessary design of a pivot cap, the top split is designed to include bushings to accommodate the rotation of the pivot cap. Careful attention must be made with the location of the pivot cap bushings. Incorrect location will result in an improper closing of the pivot caps. The top split serves only one purpose, which is to allow for any cooling of the casting. Attached to the top split is the gate block, which serves as the pouring basin of the mold. This will contain the geometry of the pouring basin and a portion of the gate. As mentioned earlier, the mold boss insert (MBI) is located within the bottom split. The MBI is used to form part of the piston geometry around the exterior pin hole. Since the piston design requires the location of the oil drain notches above the pin hole, the MBI will incorporate the geometry necessary to cast these notches. Also located within the mold halves is the core pin, illustrated in figure 11.

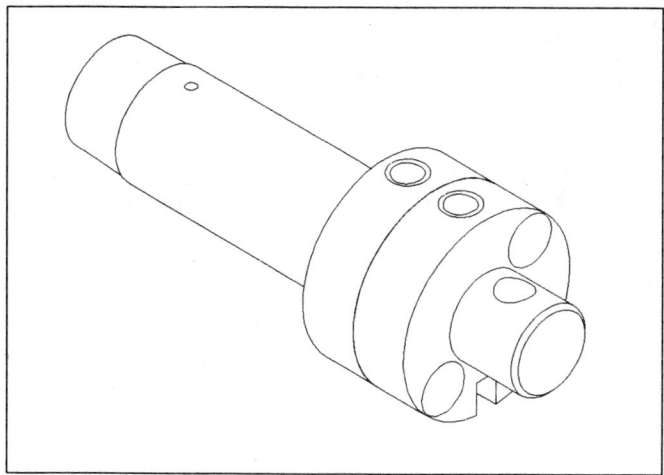

Figure 11 - Core Pin

This pin will fit through the bottom split and MBI to form the cast state of the piston pin hole. The bottom split acts as a positive stop for the travel of the core pin. For a center riser design, the diameter of the riser will vary according to the size of the cast piston allowing enough volume of metal in the riser to feed any shrinkage of the piston. At the bottom of the mold halves is a ring assembly which consists of a ring and ring insert. This ring assembly is illustrated in figure 12.

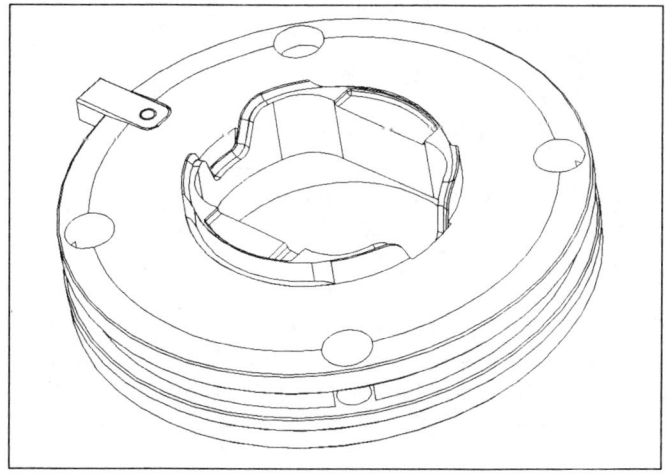

Figure 12 - Ring Assembly

The ring guides the mold halves during opening and closing of the mold. A ring insert, located inside the ring, creates the bottom geometry of the piston. The core detail is a five-piece assembly which creates all core features of the piston. It consists of a t-section, flips and sides blocks. The core detail is shown in figure 13 and 14.

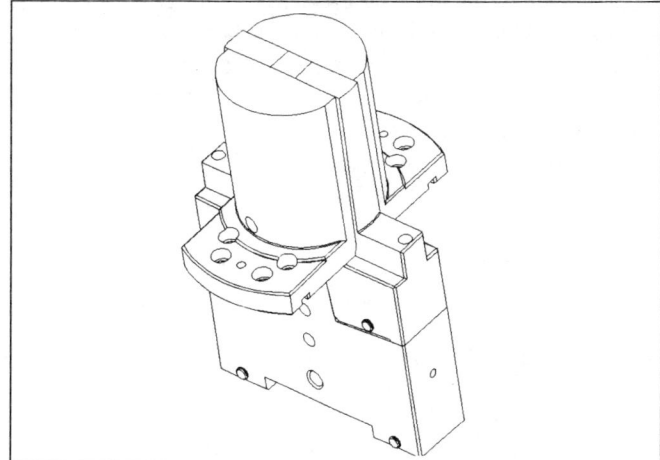

Figure 13 - Core Detail

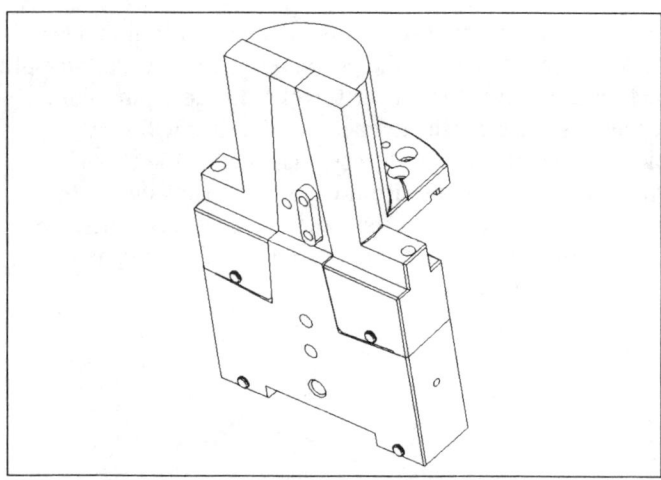

Figure 14 - Core Detail with Side Block Removed

Each item will contain some core geometry that, when assembled, will form the entire interior geometry of the piston. The core detail is designed with five (5) different parts so the casting can be removed properly from the mold. The collapsing of the core works in the following way:

1. T-section drops down from the mold
2. Flips drop down with the t-section because of the 5-8 degree angle between them.
3. One side block will move towards the center of the piston and then drop down from the mold.
4. The second side block moves towards the center of the piston.

Due to the opening procedure of the core, the removal of the casting from the core detail is permissible. Other components of the mold are considered standard parts. All of these parts are located in a separate computer library so that when designing a new mold they will be automatically placed within the new design. These standard parts consist of the following:

1. Flip Keys
2. Side Keys
3. Spring Bushings
4. Spring Plug
5. Fountain Plug
6. Clevis Pins
7. Cotter Pins
8. Rex Inserts
9. Indicator Notch Insert

Since the size and shape of these items will never change, there is no need to create a 3-D model therefore greatly reducing the design time. The next step in designing the mold is to transfer all piston geometry to its necessary components of the mold. With the use of an additional module of Pro/Engineer for mold design, all piston geometry can be transferred to the corresponding mold component while maintaining all associativity between the piston and its mold. The mold module takes the piston geometry and makes a reference part, with a defined shrink factor, of the piston design. A reference part, with shrinkage, is made for every mold component that will contain any part geometry. The easiest of all mold components to transfer piston geometry to is the mold boss insert (MBI). The procedure begins by assembling the reference part with the MBI. A process called "trim" will transfer all geometry to the MBI that comes in contact with the piston geometry. Figure 15 shows piston geometry already processed on the MBI.

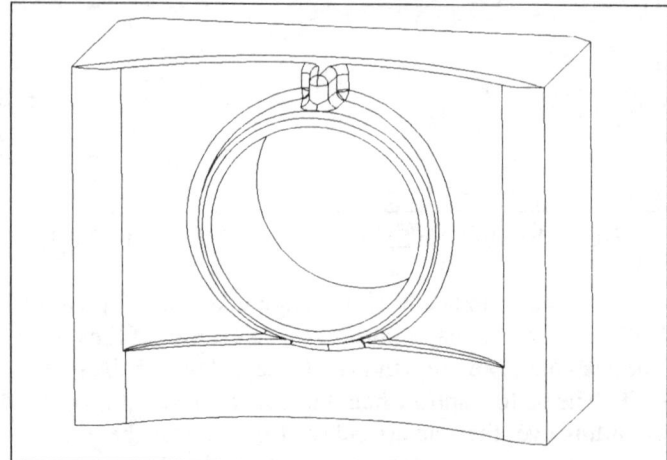

Figure 15 - Processed Mold Boss Insert

Any other features, such as creating the hole to accommodate the core pin, are then performed to the MBI. Since the ring insert must mate up accurately with the core in order to create a good seal, it becomes more complicated than the MBI. In a process the same as the MBI, the piston geometry is transferred to the ring insert. Figure 16 and 17 show the ring insert before and after part geometry is processed on the component.

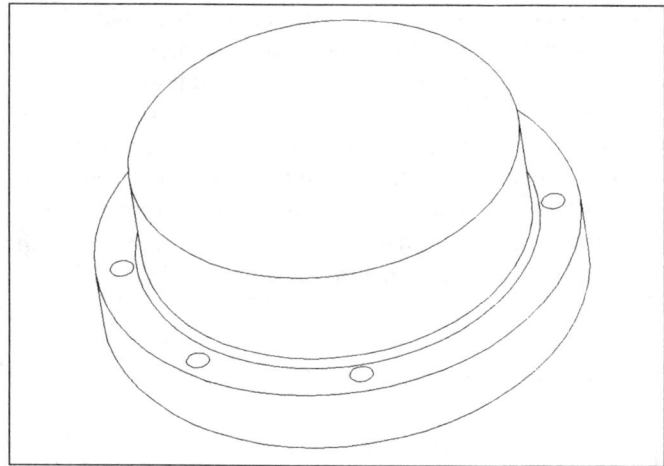

Figure 16 - Ring Insert Before Geometry

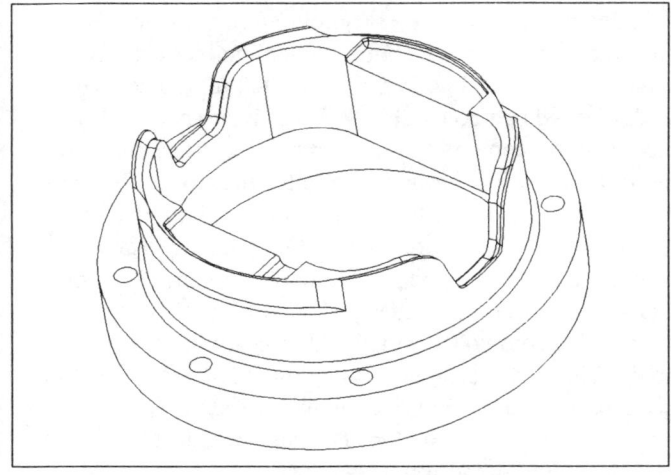

Figure 17 - Ring Insert With Geometry

Once the piston geometry is transferred, features are added (removed) to provide the seal between the ring insert and core. These features must be created accurately so that there is approximately .005" clearance between the ring insert component and the core component. The core component contains the most complex geometry of all mold components. Because of the complexity of the five-piece core detail and the limitations of the computer hardware, a slightly different approach is used. Instead of using the entire five-piece core detail to perform the trim feature on, a model called "core dummy" is used. The core dummy represents the side blocks, flips and t-section of the core detail but is only one component compared to five (core detail). Figure 18 illustrates the core dummy before part geometry.

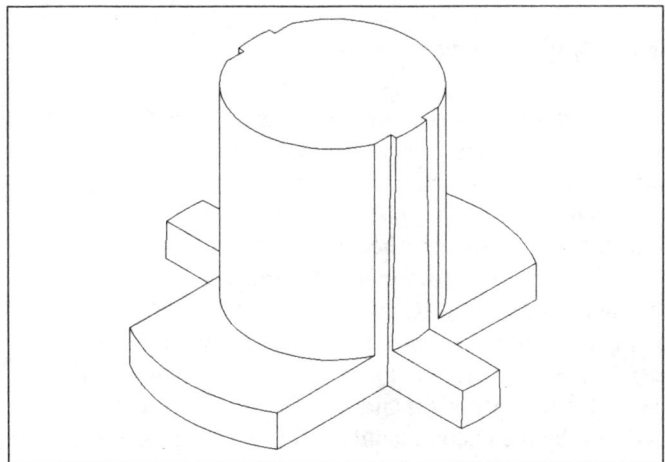

Figure 18 - Core Dummy

If a change is made to the core detail, that change is also reflected in the core dummy because of the relations set-up between the two components. The procedure to add the trim feature to the core dummy is identical to the previous ones. Features are then added (removed) to allow approximately .005" clearance between the core dummy component and the ring insert component in order to provide a good seal when pouring the metal to make the casting. Illustration 19 shows the finished core dummy.

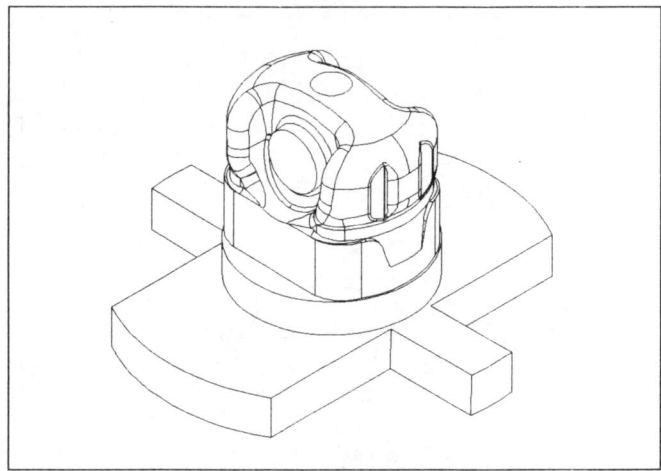

Figure 19 - Finished Core Dummy

The gate geometry, which has already been determined, is added to the mold halves by using a pre-existing 3-D gate design and performing a procedure called "cut out by reference". Using a cut out by reference will reference the gate design onto the mold halves and if a change is made to the gate design model, that change will then be reflected back into the mold halves. Illustration 20 represents the mold half with gate geometry.

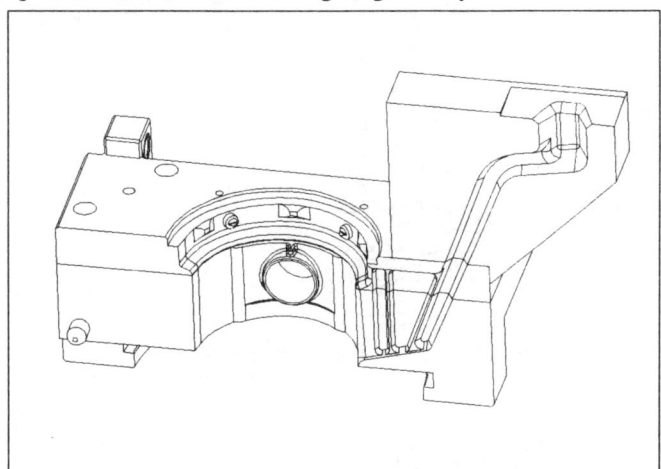

Figure 20 - Mold Half with Gate Geometry

The gate design model can be easily changed to accommodate any size piston by changing the gating scheme and the runner size if necessary. With all piston geometry transferred to the pertinent mold components, 2-D drawings of all mold components are created. These drawings are finalized and checked for approval. There are no errors with piston part geometry because of the use of the 3-D models. An assembly can be created of all mold components in its closed state and by a simple command within the software, any interference between components can be detected thus showing any problems in the mold design. A very powerful tool in finalizing a mold design before actual building of the mold. Figure 21 shows the mold half assembled with the core pin on the ring assembly.

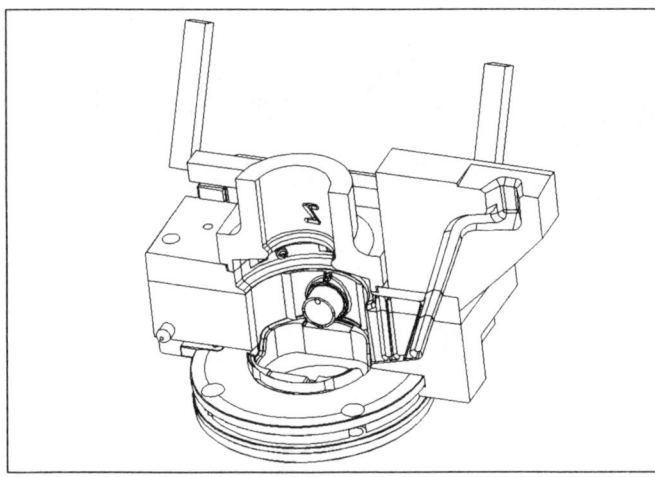

Figure 21 - Mold Half with Core Pin

Figure 22 and 23 show the complete mold half assembly with the pivot cap in the closed and open position, respectively.

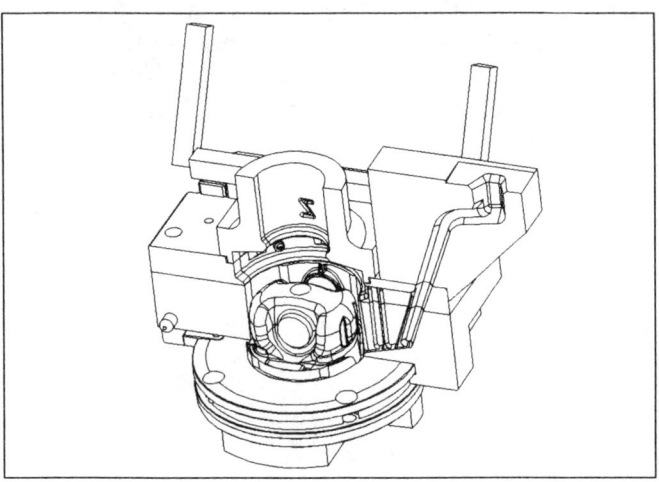

Figure 22 - Mold Half, Closed Pivot Cap

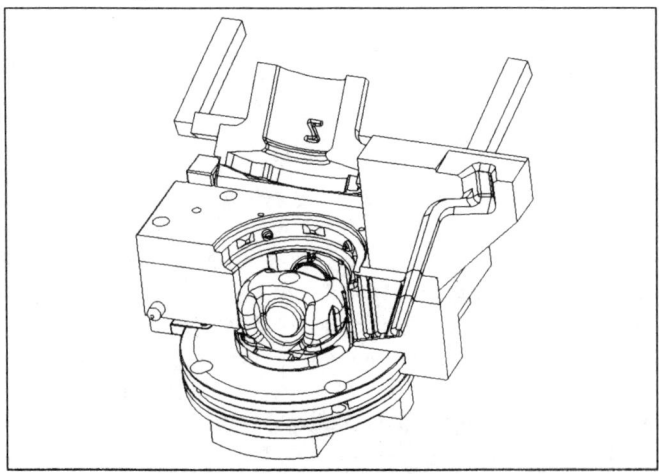

Figure 23 - Mold Half, Open Pivot Cap

Building of the mold is done by a respectable tool builder. This specialized tool builder is very experienced in building gravity molds using computer generated part geometry. They have been working in conjunction with the piston manufacturer since the piston manufacturer first implemented their new CAD/CAM system. All standard components designed by the piston manufacturers are fabricated by the tool builder and placed in their inventory. When a new mold is to be built, the tool builder can save up to four weeks build time by being able to pull the desired standard mold component blank from inventory and then machine part geometry to the mold blanks. Items that are placed in inventory consist of the following: five different styles of the core detail, five different sets of bottom splits, two varieties of the top split and a standard ring. Having different styles of components allows for less metal removal when machining part geometry. Five different styles of the core detail accommodate small and large diameter pistons and by changing the angle between the mating surface of the t-section and flips permits a core that, if necessary, can collapse quickly or gradually. The designer must recognize when designing the piston what core detail will work best for that particular mold. By having a variety of bottom splits also allows for small or large diameter pistons. The mold bore diameter of the bottom splits varies in size so that metal removal is kept to a minimum when performing the machining. The top splits are simple to machine therefore only two different styles are in inventory. The geometry of the ring is standard. Therefore they have been pre-machined except for the diameter that houses the ring insert. Once this diameter is determined, the ring insert diameter is then machined on the ring. The ring inserts start out as blanks with only bolt and dowel holes machined. All items mentioned under STANDARD PARTS OF ALL MOLDS have already been fabricated and placed in inventory in mass quantities. By setting up standard parts of all the molds and placing them in inventory one can see how the building of the mold has been drastically reduced.

MOLD MANUFACTURING

Transferring of part geometry to the tool builder begins at the piston manufacturer with the creation of IGES files (initial graphics exchange system). IGES file of all mold components that contain piston and gate geometry are created for transferring. An interface module exports part geometry to an IGES file, which contains all surface geometry of that particular model. All IGES files created for the mold components are then copied to a 3 1/2" floppy disk which is then sent, along with mold prints, to the tool builder. They receive the floppy disk(s) and transfer the IGES files generated by the piston manufacturer into the proper software so that machining programs can be written. The tool builder uses a computer software program to produce the necessary machining parameters and toolpaths. Before any programs are generated, the machine routing of the part is first laid out in order to see the process that is required for that particular part. Tool parameters are established such as tool size, spindle speed, feed rate, etc. The software has the capabilities to verify any radius or diameter that is unknown and to show the biggest tool that can be used in machining any given area. Programs are then generated, sometimes at night because of length of the program and to take advantage of the computer during off hours, using the defined tool parameters. Figure 24 illustrates a toolpath being generated.

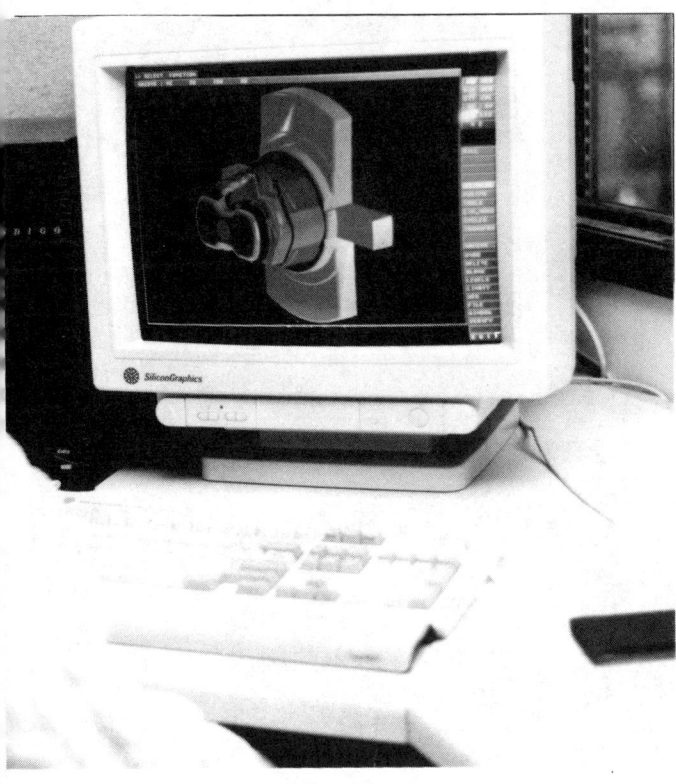

Figure 24 - Computerized Toolpath Generation

Figure 25 - CNC Milling of Core Geometry

After the program is generated for the part, post processing is performed to convert the program into machine language so that it can be downloaded to the CNC machines. Direct Numerical Control (DNC) is used to transfer the post processed program directly to the CNC machines. DNC eliminates the possibility of human error if the program would have been punched by hand. Also, DNC allows only the programmer access to the program if any changes are required, therefore only one person can change that program. The time it takes the tool builder to generate programs is one to two weeks with the core geometry being the first program generated. All programs are initially executed in wax to check for any errors. Since the core geometry program is the most complicated and time consuming of all machining programs, it is executed first. All other programs can be generated in the same time frame it takes to verify the core program. Once verified, the actual machining of the mold is performed. Figure 25 shows an actual core being milled by CNC.

All geometry, except for the top, are machined on the side blocks which are then assembled with the flips and t-section with a .015" shim placed under the flips and a .025" shim placed under the side blocks. The flips are then machined, creating a smooth blend between the flips and side blocks. The last machined feature of the core is the top core geometry. With all machining done to the core the shims are then removed creating a parting line step between the side blocks, flips and t-section which is necessary for the parting of the core after the piston has been cast. The tool builder is implementing an integrated index arm that will allow the machining of the entire core to be done in one assembly with only one program. The tool builder will use other means of machining if they feel it is not feasible to toolpath the mold component. At times they have been known to use Electrical Discharge Machining (EDM) for the ring insert. It is easier to produce the reverse blank of the actual mold component and burn out the necessary piston geometry. All other mold components are machined using the programs generated by the computer. The mold is fully assembled and ready for any bench work that needs to be done. Bench work consists of stamping the mold components with their names along with the mold's piston number. Sand blasting is done to the mold to remove any unnecessary sharp edges. The mold is then sent for final inspection to check all pertinent dimensions against the mold drawings. Once approved, the mold is delivered to the piston manufacturer.

PISTON PROTOTYPE PRODUCTION

Over the 80 plus years that Zollner has been producing pistons, manufacturing procedures for prototype pistons have been refined such that an entire division of the company supports prototype piston production. This production involves two separate processes. One being prototype casting and the other being prototype machining. The prototype casting cell is comprised of mold machines, aluminum alloy crucibles and alfin bonding equipment such that small batches of prototype pistons can quickly and efficiently be processed. A new prototype mold, once received from the mold manufacturer, is assembled onto the mold machine, mold coating is added for proper casting solidification and sample piston castings are poured. Illustrated in figure 26 is the prototype casting cell.

Figure 26 - Prototype Casting Cell

Figure 27 - Prototype CNC Piston Machining

These castings are then dimensionally checked in a quality assurance audit using the latest Coordinate Measurement Machine technology. X-ray equipment is also employed to determine the soundness of the piston casting. If a problem occurs during the audit, the piston mold is then modified to correct the problem. With the use of the new mold manufacturing process, mold dimensional problems are almost non-existent. This has drastically cut the amount of time spent reworking incorrect piston molds. Once the pistons have successfully passed the quality assurance audit, actual piston castings are produced.

The piston castings are then passed onto the prototype machining center. The piston machining process is performed using the latest Computer Numerical Control turning and milling technology available. Skirt and land configurations are produced using fully programmable CNC lathes without using hard tooling, which allows for quick and flexible piston manufacturing. Figure 27 shows an actual CNC machining operation of a prototype piston casting.

Standard tooling is employed as much as possible to minimize prototype machining time. CNC toolpathing technology, very similar to that used in the mold manufacturing process, is employed for all piston top features. After the pistons are completely machined, the machined features are dimensionally checked in another quality assurance audit. Once the pistons are within blueprint specifications, they are then shipped to the customer for use in engine tests.

CONCLUSION

With the aid of new CAD/CAM technology and the use of standard pre-designed and pre-manufactured mold components, Zollner Corporation has been able to reduce the time of design-to-manufacturing of a piston and its mold from twenty-one weeks down to twelve weeks (a 57% decrease). By reducing the design-to-manufacturing time, their customer base will continue to grow thus establishing Zollner Corporation as a continuing leader in the piston market. With the technology Zollner Corporation has incorporated into its engineering department, they have achieved a significant step in the race for excellence.

ACKNOWLEDGMENTS

The authors would like to thank the following people for their technical assistance in achieving this novel piston rapid prototyping procedure: Don Firestine, Gerald Roberts, Neal Shady and Paul Brinker of Zollner Corporation and Rob Marr, Dennis Lloyd and Fred Willis of C & A Tool Engineering, Inc.

950525

Isuzu New V8 - V12 PE-Series Diesel Engines

Jiro Saito, Masaru Odajima, and Tetsuya Harada
Isuzu Motors, Ltd.

ABSTRACT

"ISUZU GIGA" series of new heavy duty commercial vehicles introduced in Japan in November 1994 are equipped with new V-type, naturally aspirated PE series engines. The PE series engines supersede PD series which has been well accepted in the markets domestic and abroad. The new series has been designed to meet Japan's new 1994 exhaust emission regulations while achieving high output power, low weight, good fuel economy, and high reliability.

One outstanding feature of the PE engines is the use of dry liners which few other V engines have. This has made it possible to accommodate increase bore size and increase engine displacement by 14 % while maintaining the relatively short bore pitch of 146 mm. PE engine is in fact a high output engine with the smallest weight in class. The series includes V8, V10 and V12 engines ranging in output from 210 kW to 331 kW, the latter being Isuzu's most powerful engine. Six models differing in output power are available.

This paper is an overview of the major features of the new PE series engines, with some new innovated technological advances.

MARKET ENVIRONMENTS

EMISSION REGULATION - Fig.1 shows the history of Japan's NOx and somke emission regulation. Major points of the current regulation which took effect in 1994 are as follows:

1. Regulation of particulate matter was introduced.
2. Test pattern was changed from 6 mode to 13 mode.
3. The basis of pollutants control (NOx, CO, HC and PM) was change from ppm to weights.

Fig. 2 shows NOx and PM regulation and Fig. 3 shows test patterns.

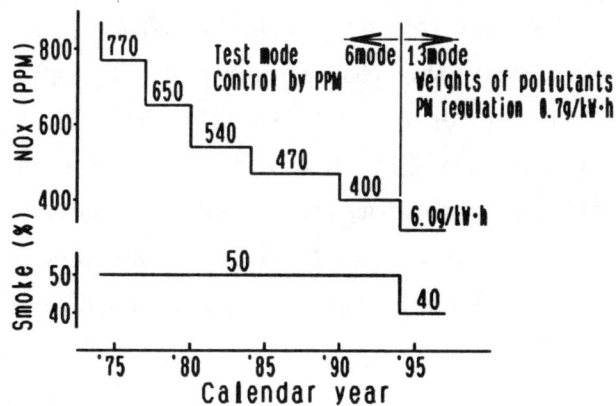

Fig. 1   History of exhaust emission regulation in Japan
(For DI heavy duty diesel engine)

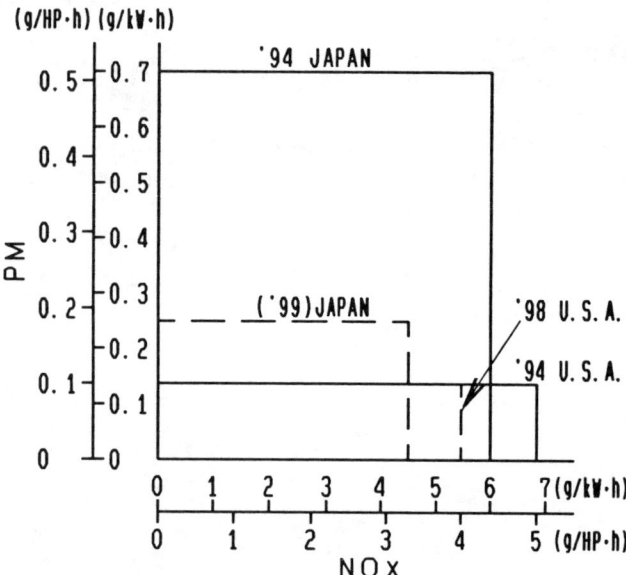

Fig. 2 NOx and PM regulation in U.S.A. and Japan
(For DI heavy duty diesel engine)

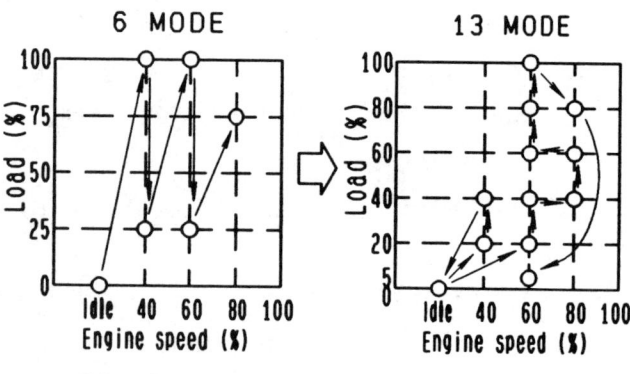

Fig. 3 Test pattern in Japan

WEIGHT CONTROL – Vehicle weight control environments have changed drastically in Japan as the gross vehicle weight limit was raised in November 1993 (Fig. 4) and, on the other hand, a stricter overloading control was enforced in May 1994. This situation inevitably translates into growing pressure for saving even more engine weight.

TREND IN ENGINE OUTPUT, DURABILITY – Fig. 5 shows that engine power has grown steadily over the past years. This trend is likely to stay with the relief of gross vehicle weight control in effect and as people require still easier drive with less fatigue.

As vehicle use rate and vehicle travel distance per year increases, demand for longer service life is likely to increase (Fig. 6).

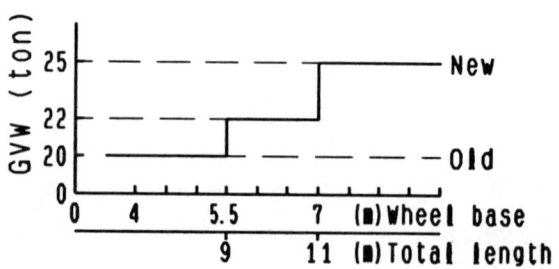

Fig. 4 GVW regulation in Japan

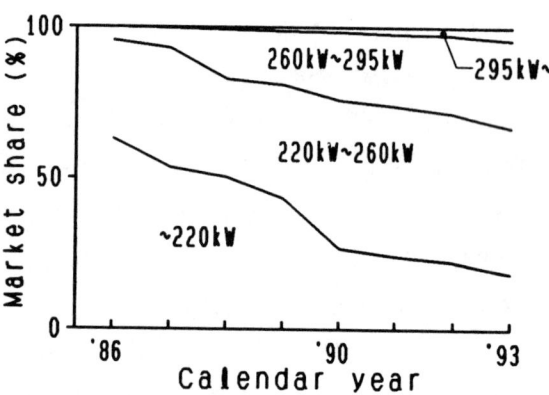

Fig. 5 Progress in out-put power for heavy duty

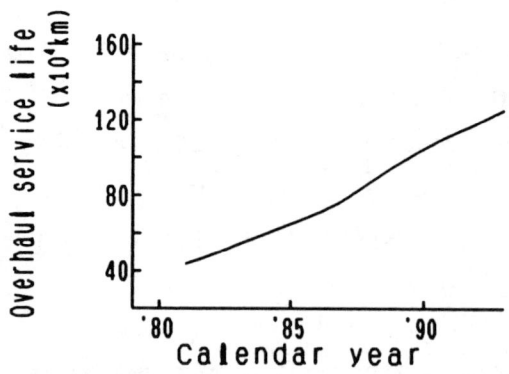

Fig. 6 Trend of engine life for heavy duty

TREND IN TURBOCHARGED ENGINES - Turbocharged heavy duty commercial vehicles have grown in population as the freeway network has expanded in Japan. Even so, there is still strong demand for naturally-aspirated engines for some custom equipped vehicles such as dump trucks as well as cargo carrying trucks (Fig. 7).

PE engines have been developed in view of these legislative moves and trend in product requirements. Fig. 8 indicates a simplified history of P series engines and Fig. 9 shows the external appearance of 12PE1 engine.

Fig. 9 12PE1 Engine

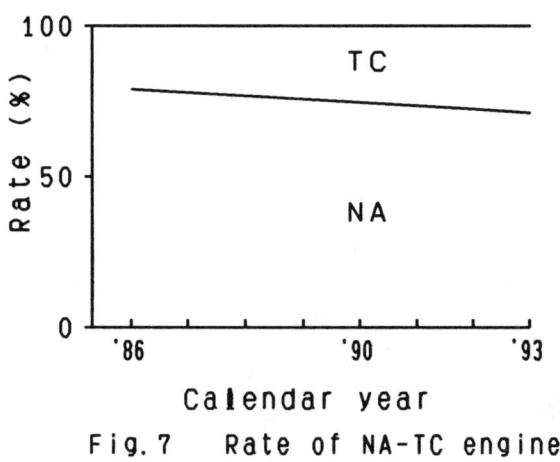

Fig. 7 Rate of NA-TC engine

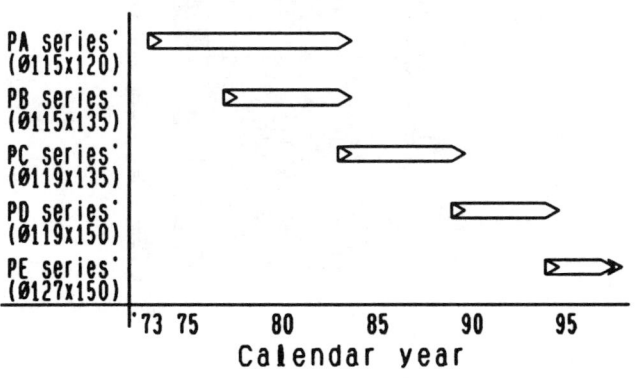

Fig. 8 History of P-Series engine

DESIGN CRITERIA

Major objectives in developing PE engines are as follows:

1. To meet 1994 Japanese exhaust emission regulations
2. To achieve optimum trade-off between high output and low fuel consumption
3. Weight saving, compact size and low noise
4. Drastic reduction in smoke emission
5. Better reliability/durability, good serviceability
6. Parts commonization within the series

To achieve these objectives, we incorporated a variety of new technology into the combustion system, injection system, etc., one of which is a dry liner type cylinder block. Table. 1 shows the main specifications and Fig. 10 is the cross section of 10PE1.

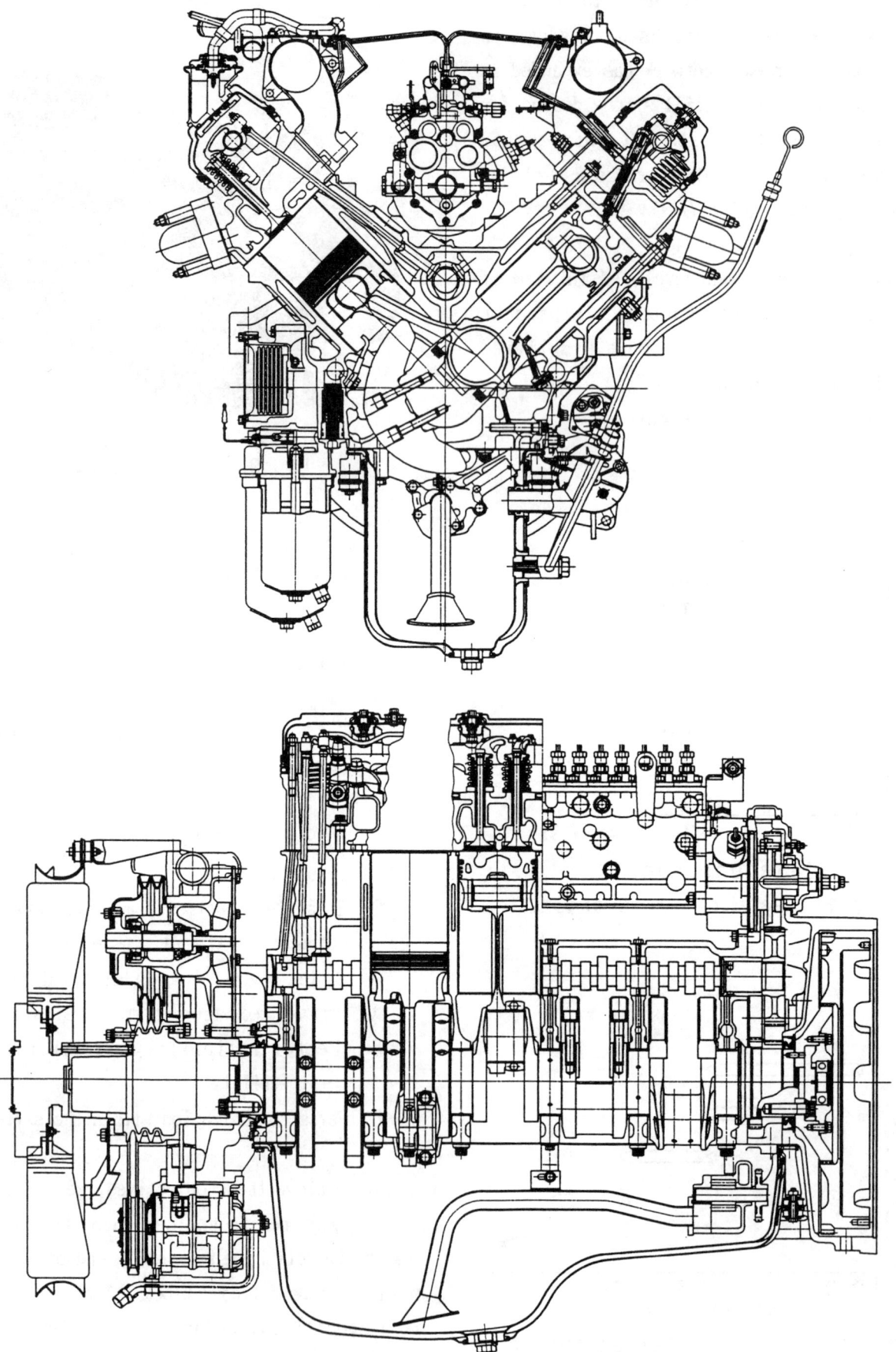

Fig. 10 10PE1 engine cross section

Table 1. Engine specification

ENGINE MODEL	12PE1-S	12PE1-C	12PE1-N	10PE1-S	10PE1-N	8PE1-S
CONFIGURATION	V12 DI			V10 DI		V8 DI
ASPIRATION	NA			NA		NA
BORE×STROKE (mm)	φ127×150			φ127×150		φ127×150
DISPLACEMENT (ℓ)	22.80ℓ			19.00ℓ		15.20ℓ
MAX. OUTPUT (kw/rpm)	331/2300	309/2300	283/2300	265/2300	239/2300	210/2300
MAX. TORQUE (Nm/rpm)	1568/1400	1460/1400	1392/1400	1245/1400	1156/1400	1000/1400

DESIGN FEATURES

CYLINDER BLOCK – The 90° V-type cylinder block is made of cast iron and is integral with the crank case. Its wet liners have been changed to dry liners in an attempt to achieve bore-up while maintaining conventional bore pitch (146mm) and total cylinder block length. As a result, we achieved basic dimensions that minimize weight and size with the bore pitch to diameter ratio, which has the greatest effect on engine weight, reduced to 1.15. Fig. 11 shows the external appearance of a 12-cylinder block and Fig. 12 shows its cross section.

Fig. 11 12PE1 Cylinder block

There have been few V engines with dry liners because of difficulty in casting the cylinder block. In addition, in the case of the PE engine, attempts were made to make good use of exclusive machining equipment used for the former P series engines, and this made it impossible to provide a wide opening on the side of the cylinder block as a lot of dry liner type cylinder blocks (Fig. 13). This in turn made it even more difficult to, among other things, remove sand from inside the water jacket and to ensure wall thickness accuracy around the bores. We dealt with these problems by improving core fixing, sand stripping and many other casting related processes.

Maintaining proper cylinder block stiffness is very important in suppressing noise and

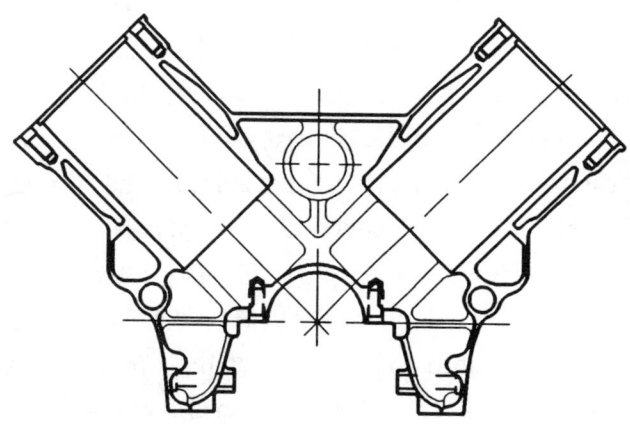

Fig. 12 Cylinder block cross section

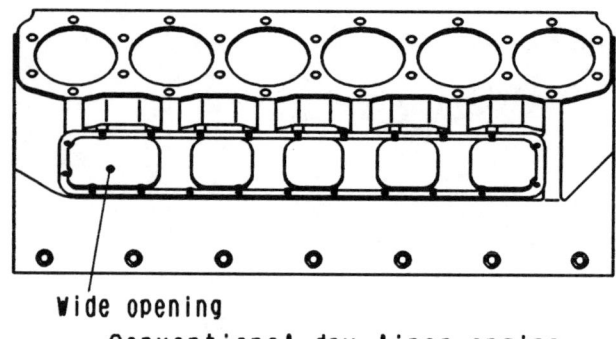

Fig. 13 Cylinder block side view

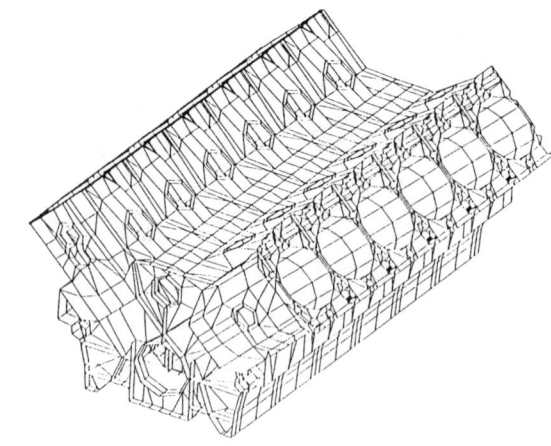

Fig. 14 Cylinder block FEM model

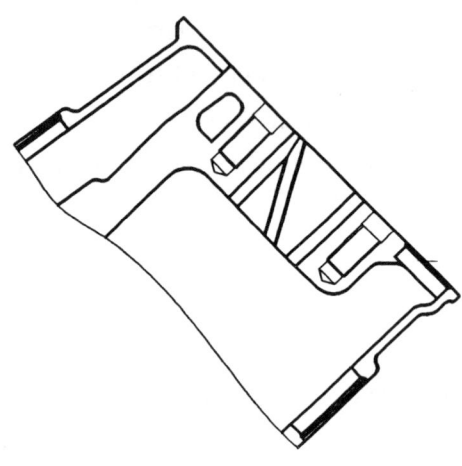

Fig. 15 Water holes between cylinder bores

vibration. In the PE engine, the aforementioned cast seal cups helped realize highly stiff basic structure. Moreover, optimum stock balance and rib layout have been achieved through modal and FEM analysis which have been in effect since early stages of development. Fig. 14 shows an example of FEM analysis.

For effective cooling of the cylinder bores, holes are drilled between bores which are exposed to especially high temperatures (Fig. 15).

The cylinder liners are made of special cast iron with thin wall (1.8 mm thick), and are fit loosely for easy insertion. For optimum fit with the cylinder block, 3-grade fitting selection is available. Cylinder block with dry liners are often vulnerable to bore distortion. This problem was solved by coupling the walls between the cylinder head bolts in straight line and this can maintain a bore distortion as well as wet liner type.

CYLINDER HEAD AND GASKETS - A patented structure where the intake and exhaust push rods are arranged side by side allowed very high degree of intake port design. Thus an ideal shape was worked out to realize significant improvement in volumetric efficiency. Fig. 16 shows how the intake port appears and Fig. 17 shows how effective it is.

To improve cooling efficiency between the valve seat and nozzle, balance of coolant flow from the cylinder block was reviewed to increase the flow rate, and the drilled hole between the valves was relocated 2 mm downward (Fig. 18).

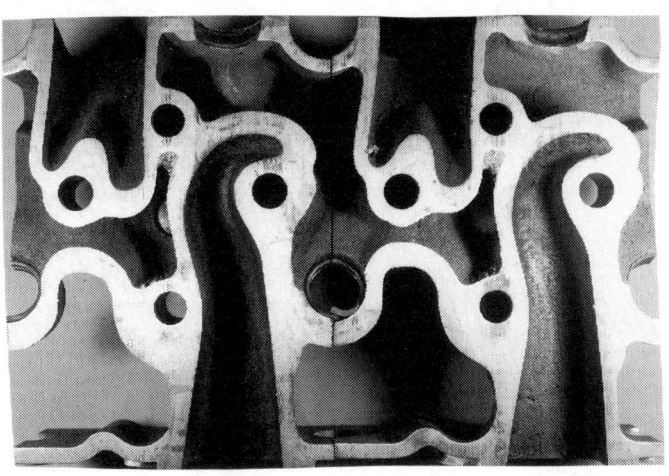

Fig. 16 Cylinder head cross section

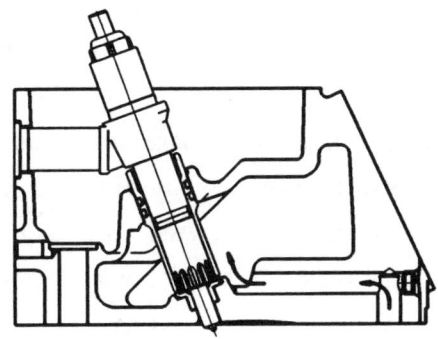

Fig. 18 Cylinder head cross section

With additional coolant passages between bores, the ribs in the cylinder head were enlarged to provide more coolant passages. These ribs, and the gasket discussed later, also helped to improve stiffness in the bottom area of the cylinder head and seal ability.

SUS steel laminated-type cylinder head gaskets, which have long been used in large quantities on conventional engines, are used on this engine, too (Fig. 19). With 5 sheets laminated, optimum surface pressure balance has been achieved. Rubber grommets made of most suitable material are used for both coolant and oil passages. These and the six fixing bolts per cylinder, which are tightened by plastic range angle method and ensure uniform axial force distribution, achieve sufficient seal.

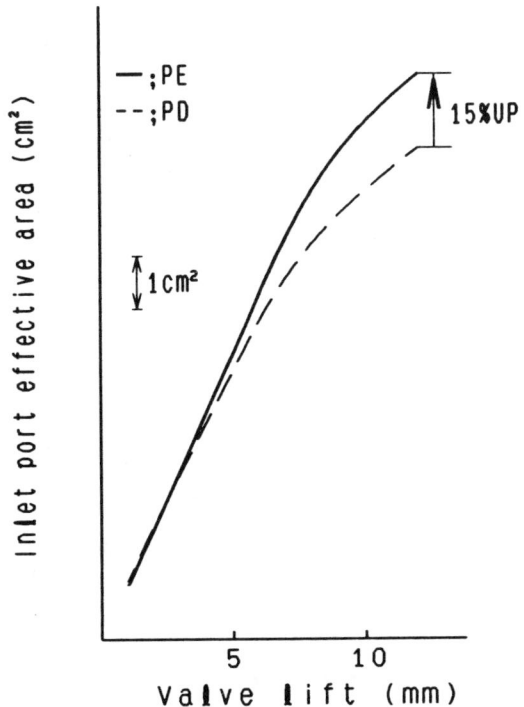

Fig. 17 Comparison of inlet port effective area

Fig. 19 12PE1 Cylinder head gasket

PISTONS AND PISTON RINGS — The piston is made of 12 % silicon aluminum. A Ni-resist ring carrier is cast into the top ring groove to minimize wear, and the position of the top ring was raised by 4.5 mm to arrange four rings over a relatively short compression height of 67.5 mm. Such an arrangement could lead to higher thermal load and thus less durability, but it was dealt with by the adoption of pistons having oil cooling channels (Fig. 20). Optimizing the oil passage size in the oil jet and the check valve resulted in a sufficient oil flow rate of 3.6 l/min at the rated engine speed, which ensures enough amount of oil fed to the cooling channels. Temperature distribution over the piston is shown in Fig. 21.

Three compression rings and one oil ring are installed on each piston. Some features of these rings are:

Top ring : Full Keystone type, barrel face
2nd ring : Inner cut type, tapered face
3rd ring : Balanced undercut type, tapered face
Oil ring : Equipped with coil spring

Fig. 22 shows the cross sections of a piston and the piston rings. The combination of the four rings, the optimization of piston land shapes for stabilizing ring behavior, and the dry cylinder liner discussed earlier all contributed to a significant reduction in oil consumption, as well as to longer service life which had been

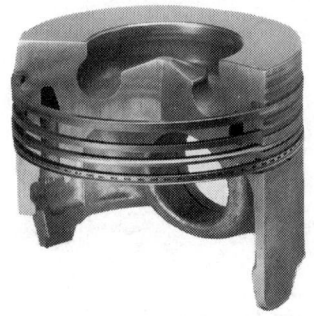

Fig. 20 Piston and rings

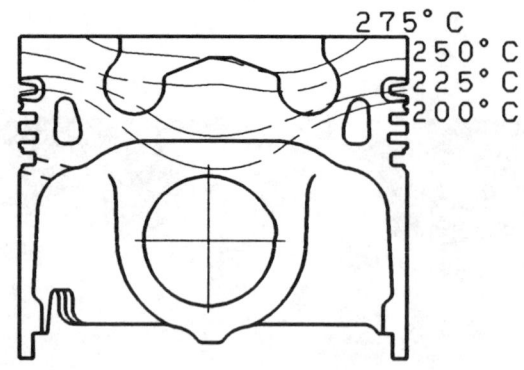

Fig. 21 Heat distribution of piston

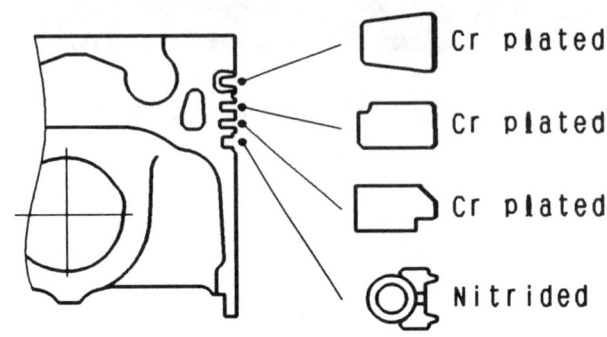

Fig. 22 Piston and ring structures

one of our major objectives. Fig. 23 shows the result of vehicle testing.

A newly designed combustion chamber, re-entrant type with center projection, is formed in the top face of each piston. The design was determined as a result of air flow analysis by simulation and of several improvements through bench test. Fig. 24 shows an example of the simulation. A ball-like combustion space formed by the side walls and the center projection generates strong squish flow and vortexes, creating ideal air-fuel mixture, which in turn helps reduce exhaust and smoke emissions simultaneously.

The stock of the piston pin was increased by 12 % to accommodate higher mechanical loads due to bore-up and to ensure sufficient stiffness.

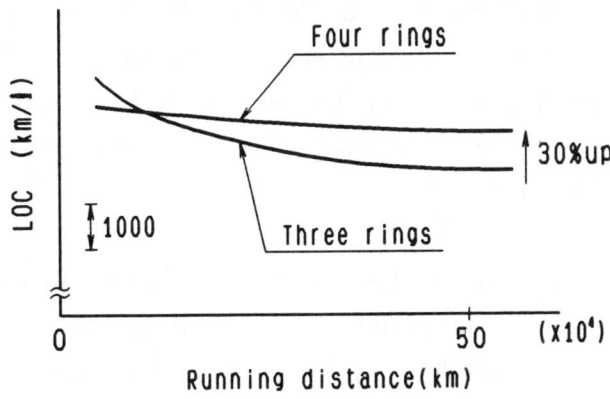

Fig. 23 Comparison of lubricating oil consumption (Vehicle test)

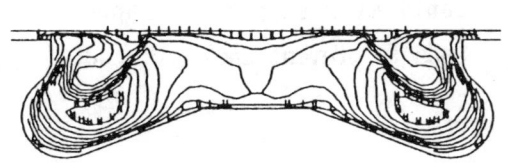

Fig. 24 Simulation within combustion chamber

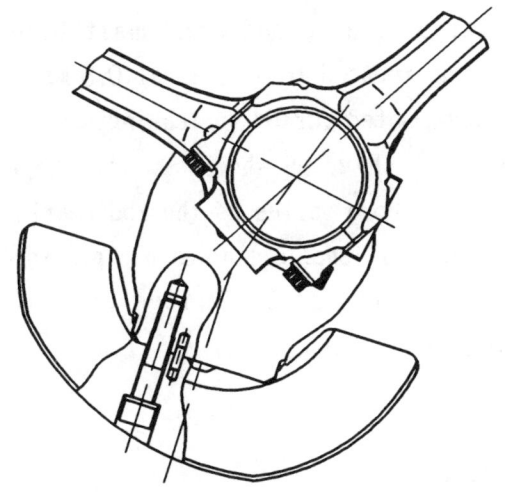

Fig. 25 10PE1 Cross section of crank barancer

Fig. 26 Comparison of oil seal

Fig. 27 Connecting rod

CRANKSHAFT AND CONNECTING ROD — The crankshaft body, softnitrided carbon steel forging, is common to that of PD series engines. Migration in unbalance quantity due to bore-up was dealt with by increasing the weight of the bolted balancers in 8PE1 and 10PE1. Positioning of the balancer was changed from conventional fit-in to knock-pin drive to improve manufacturing efficiency (Fig. 25).

Conventional redial-type oil seals at both ends of the crankshaft were replaced by axial-type oil seals integral with the slingers (Fig. 26), which helps prevent lip damage during assembling operation, and the use of fluorosilicon in the rubber material significantly reduced wear rate.

For the connecting rod, effort was made to deal with increased inertial force due to bore-up by conducting FEM analysis and measurement of stresses (Fig. 27). An example of FEM analysis is shown in Fig. 28. Since the bolt size was increased from M12 to M13, the distance between the threaded hole and crankshaft bore on the large end reduced to a critical 0.5 mm, but this was compensated for by increasing stock in axial direction. Fig. 29 shows how the shape of the crankshaft bore varies as the rod bearings close in. Two additional oil holes are provided in the tapered small end to improve piston pin and bushing lubrication.

TIMING GEAR TRAIN AND VALVE MECHANISM - The following improvements were incorporated into the timing gear train to deal with the higher pressure injection system which in turn increased driveng torques:

1. Heat treatment of some forged gears was changed from soft nitriding to salt-quenched carburization to increase hardness.
2. The support of the idle gear assembly, in which small and large idle gears are coupled together, was changed from cantilever support to suppo rt on both ends to prevent uneven contact (Fig. 30).

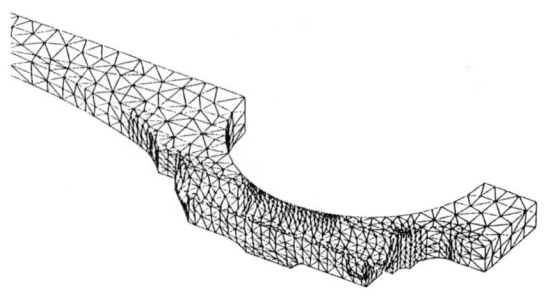

Fig. 28 Connecting rod FEM model

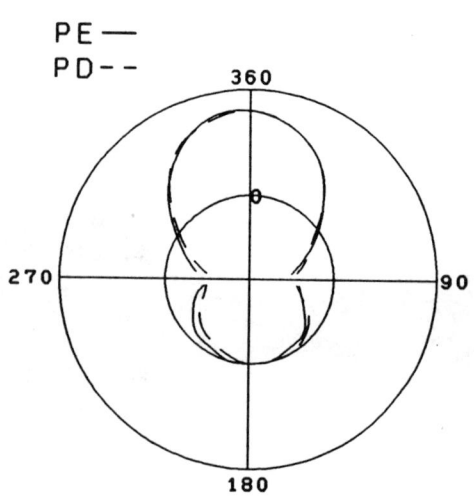

Fig. 29 Connecting rod close in

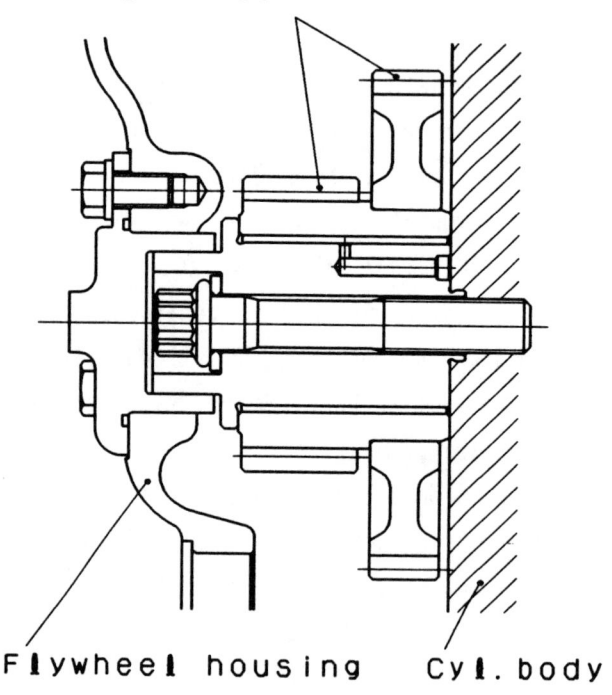

Fig. 30 Cross section of idle gears

The OHV mechanism is common to that of PD-series engines. On top of many improvements made in the past, with the introduction of the PE series, ceramic material was brazed to the friction interface of flat tappet where contact with cam to further improve wear resistance (Fig. 31, 32). For the same purpose, electronic beam hardening was applied to the contact face on the push rods.

Fig. 31 Ceramic tappet

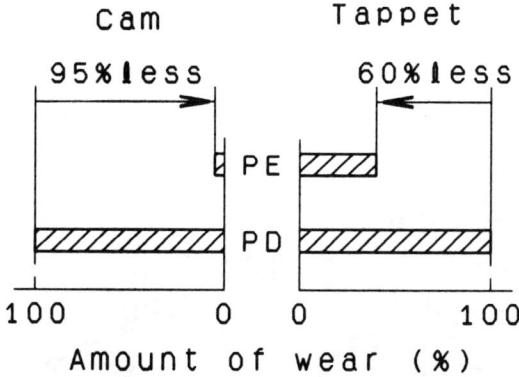

Fig. 32 Comparison of cam and tappet wear (Bench Test)

LUBRICATION AND COOLING SYSTEM - Oil pump gear tooth width was widened to increase oil discharge by 19 % and ensure sufficient lubrication for longer service life. In addition, the number of oil cooling cores was increased from 5 to 8 in order to increase heat radiation by 60%, and the temperature at which cooler bypass oil thermostat opens the oil passage was lowered by 8 %. All these actions are intended to stabilize oil temperatures and improve durability.

In the cooling system, the water pump body was changed from iron casting to aluminum casting, which resulted in 33 % weight saving. A one-piece unitized-type mechanical seal was used to improve reliability, and its carbon material was improved to withstand non-amine-base LLC.

INLET AND EXHAUST SYSTEM - The crosssectional area of the inlet manifold was increased by 34 % to ensure enough intake air for increased displacement. A resonator made of plastic material is provided on the secondary side of the inlet system. This arrangement, shown in Fig. 33, is intended to reduce intake noise which could add up to cab noise.

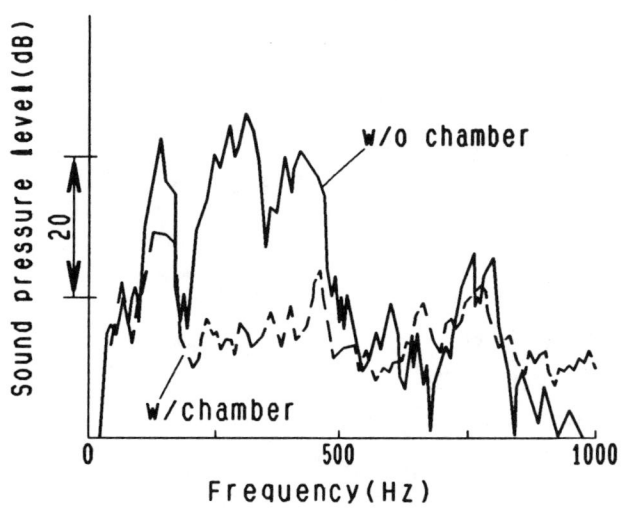

Fig. 33 Effect of inlet resonance chamber

The vermicular cast iron exhaust manifold is of split type. Careful simulation was carried out to determine stock distribution and rib arrangement in order to prevent gas leak due to thermal crack or distortion. In addition, optimum combination of circular and oval fixing holes was selected to allow for thermal expansion. Fig. 34 shows an example of simulation.

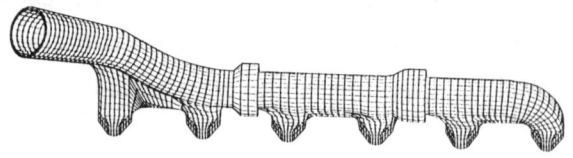

Fig. 34 Exh manifold simulation model

FUEL SYSTEM - We took the following approach regarding the fuel system on the PE engine:

First, the injection pressures, injection timing and diameters of the injection nozzle holes which are required to achieve the targets of emission levels, output, fuel consumption, and exhaust temperatures were studied in laboratory testing. Then fuel systems that can achieve those goals with minimum cost were selected. The results of our work are shown in Table 2. The fuel systems differ on 12PE1-S and other engines.

Table 2. Fuel system specification

ENGINE MODEL	12PE1-S	OTHERS
INJ PUMP	TICS	S3SS
DELIVERY VALVE	C.P.V.	C.P.V.
GOVERNOR	ELEC.	RLD-J
AUTO TIMER	—	ELEC.
INJ NOZZLE OPEN PRESS. (MPa)	15.7~22.1 (Two stage)	
NOZZLE DIA. (mm) & NUMBER	φ0.21 × 7	

12PE1-S, which has the maximum output in the series, is equipped with Zexel's high pressure in-line HD-TICS pump, which is already introduced in SAE paper (1) (Fig. 35). This pump had already been on the market installed on 6 in-line engines such as our 6WA1TC engine (2), but the application of this pump to 12-cylinder engine realized only after many modifications to improve reliability. At the same time, a constant pressure valve (CPV) was employed to improve injection waveforms in order, especially, to reduce smoke emission. In addition, an electronic control governor was employed to compensate for the tendency of fuel injection being inadequate at high speeds, which is one drawback of CPV, and also to ensure proper amount of injection at all speed ranges.

On the other five engine models in the PE series, that is from 12PE1-C to 8PE1-S, Zexel S3SS pump is used. With the casting for S7S used for the pump housing, machining used on S3S was applied to this pump to improve stiffness. S7S and HD-TICS parts are also used in the vicinity of the tappets to ensure proper durability. In addition, the pump for these engines is equipped with the same CPV as one used in HD-TICS to

Fig. 35 12HD-TICS pump

improve the waveforms. For the mechanical governor, the all-speed-type RLD-J governor is employed so that a large amount of positive angleich value can be ensured. The auto-timer is an electronically controlled hydraulic timer which conducts optimum control of injection timing to achieve significant improvement in fuel economy, startability, white smoke emission and smell.

As on PD engines, the injection nozzle opens in two steps at different pressures (Table 2), which contributes to prevent "Car-knock" and reduce idle noise. As a result of matching with the injection pump and combustion chamber discussed above, the size of injection hole was determined to be 0.21 mm in diameter, with seven holes per nozzle (Fig. 36). Because of the extremely small hole size, an edge filter is installed at the inlet of the nozzle holder to prevent the holes from being clogged.

For all PE engines, dual SUS injection pipes are used to ensure reliability.

ELECTRONIC CONTROL SYSTEM - The following devices are electronically controlled:

1. HD-TICS
2. Electronically controlled governor (HD-TICS pump)
3. Electronically controlled timer (S3SS pump)

As used on our engines, reliability of these devices are now fully proven in the field (2)(3)(4).

PERFORMANCE

EXHAUST GASES - The injection timing is retarded from 8° BTDC on PD engine to 5° BTDC on 12PE1-S and to 4° BTDC on other engines in the PE series. As a result of the aforementioned employment of the high pressure injection system and improvement of the combustion system including the new combustion chamber, PM and NOx emissions dropped to the levels shown in Fig. 37, which are below the regulatory standards. CO_2 and HC also meet the standard with comfortable safety margin. One major objective in developing the PE engines was to reduce smoke emission to a level people perceive as acceptable. Now smoke is nearly invisible either at start, acceleration or in full load operation, and is of course far below the

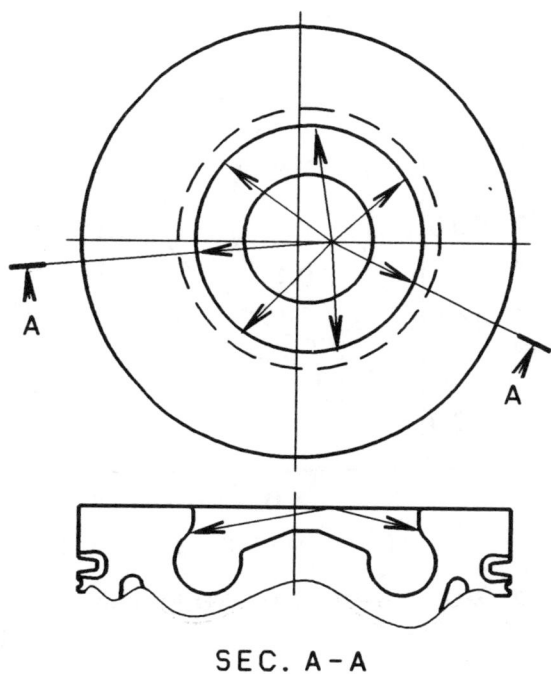

Fig. 36 Injection direction of 7 nozzle holes

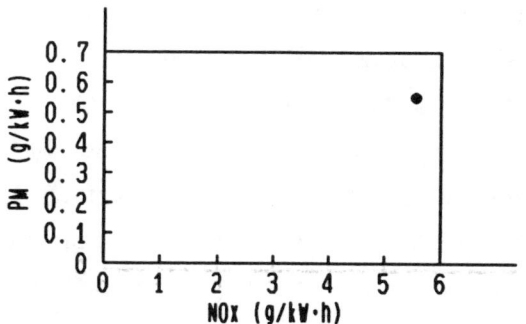

Fig. 37 PE engine test result

mandatory standard. This will mean a significant improvement in customer acceptance. Fig. 38 shows comparison of smoke emissions under full load.

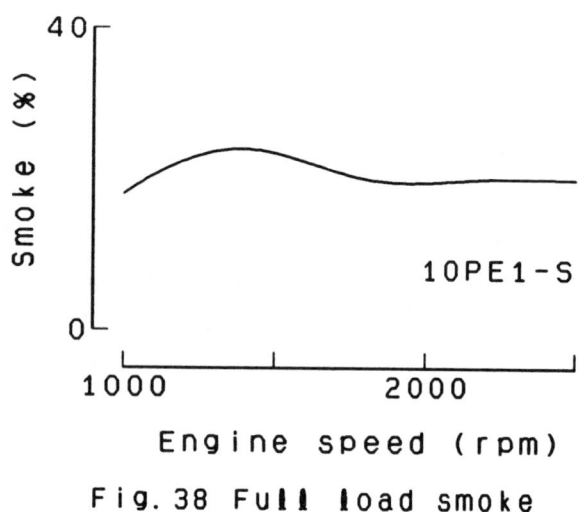

Fig. 38 Full load smoke

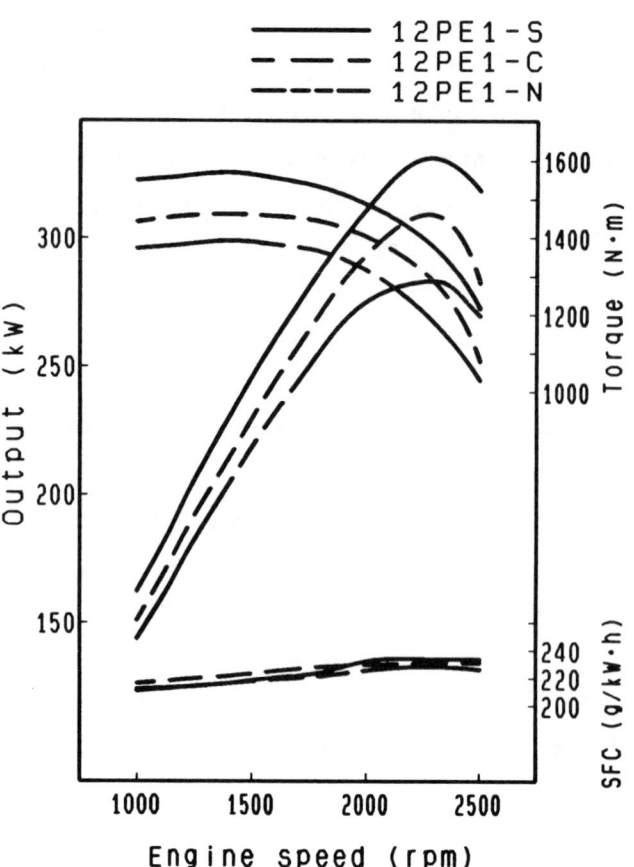

Fig. 39 12PE1 Full load performance

OUTPUT POWER AND FUEL ECONOMY - The full load performance curve of 12PE1 is shown in Fig. 39. Dump trucks and custom equipped urban trucks require good driveability which in turn requires high torques from the low speed range up. As shown in Fig. 39, the PE engine achieves a flat torque over all the speed ranges to enable vehicle operation free from fatigue. The rated output of the 12PE1-S engine, 331kW, is by far the greatest among trucks (non-tractor-trailers) in Japan.

Stricter exhaust emission regulation normally adversely affects fuel economy. However, aforementioned improvements to the injection and combustion systems realized about the same full-load minimum fuel consumption as that of PD engines, and in addition improvement to the fuel consumption map achieved fuel consumption during vehicle operation even lower than that of our conventional vehicles. Fig. 40 shows the fuel consumption map.

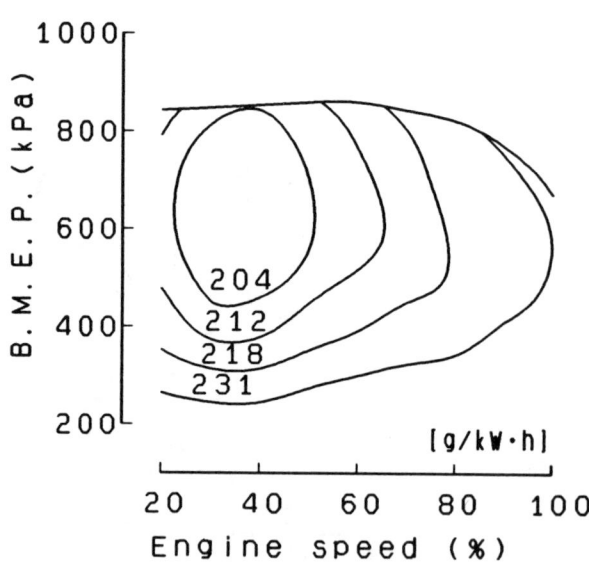

Fig. 40 Fuel consumption map (12PE1-S)

Moreover, engine startability has been significantly improved by increasing the compression ratio by 0.5 and providing a helix for timing advance in an injection pump plunger.

NOISE, WEIGHT - Generally, there is a tradeoff relation between noise and weight. Even so, in PE engines, both weight saving and low noise level are achieved at the same time. Efforts to reduce mechanical noises right from the base design stage by means of FEM analysis, and efforts to reduce combustion noise using combustion noise meters, and a noise insulation cover installed near the engine led to this achievement. The noise level is shown in Fig. 41 and weights are compared in Fig. 42.

RELIABILITY/DURABILITY - A total of 150 prototype engines were built for a total of 45,000-hour bench durability tests and a total value of six million km vehicle durability tests.

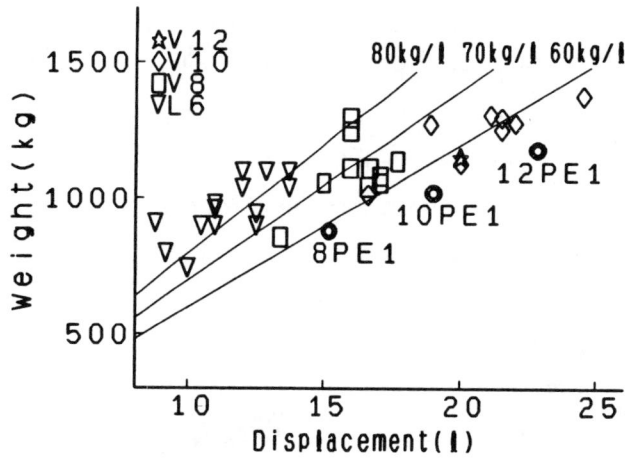

Fig. 42 Engine weight to displacement ratio

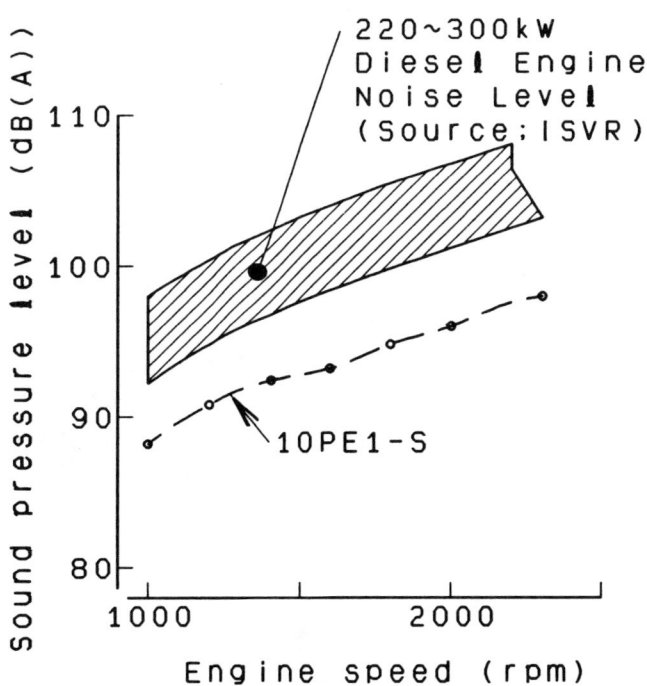

Fig. 41 Engine noise level

The bench tests were run on a large number of established driving patterns to allow for a variety of ways in which the vehicles are used in the field. New mechanisms and parts have been evaluated in laboratory tests as well under new bench test procedures. These test procedures are summed up into the following four categories:

1. Specific laboratory tests of individual components
2. Accelerated abuse tests of complete engines on dynamometer
3. Durability tests on dynamometer
4. Actual vehicle durability tests

Item 2, for example, is run under very severe conditions including excessively high speeds, overload and high coolant temperatures. Test duration was stretched and accelerated test conditions were modified to make sure that longer service life has been attained for customer satisfaction.

As a result of these efforts, we have succeeded in providing the PE series engines with sufficient service life, reliability and durability.

CONCLUSION

This paper introduces the outline of design concept for PE series engines used on Isuzu Motors' new heavy duty commercial vehicles, "GIGA". Major points in developing PE engines were the development of dry liners in the cylinder block, as well as new injection and combustion systems. A large amount of Isuzu's engineering resources, including those of production departments, were rounded up to achieve all the targets including output power, weight saving, fuel consumption and reliability.

Further efforts are planned to develope technology in order to achieve still lower exhaust and noise emissions, still better power/fuel consumption trade-off, lighter weight and greater reliability, for which user needs are likely to keep growing.

REFERENCES

(1) H.Ishiwata and H.Okubo "A new series of timing and injection rate control systems, AD-Tics and P-Tics" S.A.E. paper 880491

(2) M.Kita and M.Wakabayashi "New 12L 6WA1TC Turbocharged Diesel Engine" S.A.E paper 930718

(3) M.Wakabayashi, S.Sakata and K.Hamanaka "Isuzu's new 12.0L micro-computer controlled turbocharged diesel engine" S.A.E. paper 840510

(4) T.Hashimoto, K.Okada and T.Oikawa "Isuzu new 9.8L diesel engine with variable geometry turbocharger" S.A.E. paper 860460

950526

Piston Cooling with Shaking-Up Heat Pipes (SUHP) and Thermal Analysis of the Cooling System

Yiding Cao and Qian Wang
Florida International Univ.

ABSTRACT

An engine piston cooling method incorporating shaking-up heat pipes (SUHPs) is described. In shaking-up heat pipes, the liquid return from the condenser to the evaporator section is achieved by a high-frequency shaking-up action. In addition, the liquid splash and impingement on the inner surface facilitate temperature uniformity along the heat pipe. The concept of the SUHP is verified by experimental observation of a transparent heat pipe and thermal testing of a copper/water SUHP. A comparative thermal analysis on the SUHP and gallery cooling systems is performed. The approximate analytical results show that the piston ring groove temperature can be significantly reduced using the heat pipe cooling technology, which may contribute to an increase in engine thermal efficiency and a reduction in environmental pollution.

NOMENCLATURE

A	area, m
c_p	specific heat, J/kg-K
D_b	piston outer diameter, m
D_h	gallery hydraulic diameter, m
d	jet diameter, m
H	height of piston, m
h	heat transfer coefficient, W/m^2-K
$\overline{h}$	average heat transfer coefficient, W/m^2-K
k	thermal conductivity, W/m-K
L	length, m
$\dot{m}$	mass flow rate, kg/s
N	the number of SUHPs
Nu	Nusselt number
$\overline{Nu}$	average Nusselt number
P	perimeter, m
Pr	Prandtl number
Q	heat transfer rate, W
R	thermal resistance, K/W
Re	Reynolds number
r	radial distance, m
T	temperature, K

GREEK SYMBOLS

μ	dynamic viscosity, N-s/m^2
ν	kinematic viscosity, m^2/s
η	capture efficiency
ρ	density, kg/m^3

SUBSCRIPTS

c	condenser or cross section
e	evaporator
g	groove
h	hydraulic
hp	heat pipe
in	inlet
m	mean
n	nozzle
out	outlet
s	surface
t	total

INTRODUCTION

Piston cooling is a critical measure for achieving designed engine performance especially for heavy-duty internal combustion engines. As the engine is pushed to increased thermal efficiency, the piston is required to work in a much more elevated temperature environment. The higher charge temperature in a combustion chamber will also reduce the discharge of soot from the engine and reduce pollution to the environment (Senda et al., 1992). One of the working limits for a piston is the maximum temperature that a piston can sustain. This is especially important for aluminum alloy-based pistons, which have a pronounced temperature dependence on their mechanical properties. However, piston cooling is difficult to implement due to the reciprocal motion of the piston. A commonly used method for cooling pistons is the crankcase oil splash/mist undercrown cooling. Above a certain rating, additional oil cooling is necessary and this is required for medium and high speed engine pistons. Provision of an internal cooling gallery in conjunction with the oil jet allows a larger surface area for cooling and shorter heat flow paths as compared to undercrown cooling (Figure 1a). The gallery is arranged to run partially filled with oil to promote a cooling action and to make efficient use of the oil. The oil may be supplied from a standing jet fixed in the crankcase or through a drilled connecting rod. However, the gallery has a reputation of causing stress concentration and reducing piston strength in the ring groove region. Also, since the galleries are located in the upper section of the piston, it is not easy for the cooling oil to reach the gallery from the crankcase. In addition, accurate jet alignment and capture efficiency are not without problems due to the rapid reciprocating motion of the piston.

PISTON COOLING USING SHAKING-UP HEAT PIPES

Shaking-up heat pipes and an engine piston incorporating shaking-up heat pipes were proposed by Cao and Wang (1994). Figure 1b schematically shows a shaking-up heat pipe cooled engine piston. A number of shaking-up heat pipes are inserted into the piston in the region close to the top ring groove, and are arranged circumferentially. The shaking-up heat pipes can be arranged with the same pitch along the circumference to provide uniformity in the circumferential temperature. In the lower skirt region of the piston, however, the heat pipes can be placed only in certain segments along the circumference due to the existence of the piston bearings for the wrist pin joint, and the clearance needed for oscillation of the connecting rod. As a result, both symmetric and non-symmetric stud-shaped heat pipes were used (Wang and Cao, 1994). Since the heat pipe has an extremely high thermal conductance, excessive heat in the top ring region can be transferred to the lower piston section, where the cooling oil is much more accessible, and the heat can then be dissipated via splash/mist or jet impingement cooling. As a result, the temperature in the top ring region and the top skirt regions can be significantly decreased, and the piston can work at a better thermal condition. Degradation of the lubricant and the aluminum alloy will also be greatly reduced, and hence, both the performance of the piston assembly and the life of the piston/cylinder interface will be significantly improved. Heat pipes are very efficient heat transfer devices with an effective thermal conductance as high as hundreds of times that of copper. Detailed descriptions on heat pipes, including two-phase closed thermosyphons, can be found in Chi (1976) and Dunn and Reay (1982). For traditional heat pipes, liquid condensate is returned from the condenser section to the evaporator section via the capillary pumping force for the wicked heat pipe, gravity assistance for the two-phase closed thermosyphon, and centrifugal force for the rotating heat pipe. The shaking-up heat pipe may be similar in structure to the wickless two-phase closed thermosyphon, which includes an air evacuated container, and an amount of working fluid charged inside the heat pipe. However, the working principles of the shaking-up heat pipes are significantly different from those heat pipes. For the shaking-up heat pipe, the liquid return is accomplished because of the high frequency shaking-up action due to the reciprocating motion of the heat pipe. The liquid splash and impingement on the inner surface also facilitates temperature uniformity along the heat pipe. Figure 2 is a schematic representation of a straight shaking-up heat pipe. For the working condition shown in the figure, the upper section of the heat pipe functions as an evaporator, and the lower section of the heat pipe functions as a condenser. In the evaporator section, heat is conducted through the heat pipe container wall to the interior surface, where it is absorbed due to evaporation of the liquid. The vapor flows from the upper evaporator section down to the lower condenser section and condenses on the interior surface in the condenser section. The latent heat released due to the condensation is conducted through the container wall to the exterior surface where the heat is carried away by the cooling medium. The liquid condensate is returned by the inertia force and impingement due to the reciprocating motion of the heat pipe. The liquid inside the shaking-up heat pipe is dispersed and distributed over the interior surface of the heat pipe, which secures a liquid supply for the vaporization in the evaporator section. The aforementioned evaporation/condensation is a typical two-phase heat transfer mechanism. However, even without evaporation/condensation in the shaking-up heat pipe, the heat transfer is still very effective from the single-phase heat transfer point of view. Due to the high frequency reciprocating shaking-up action of the heat pipe, liquid particles will alternatively impinge upon upper-hotter and lower-colder interior wall surfaces of the container, therewith effectively carrying the heat from the upper section to the lower section. The SUHP shown in Fig. 2 is a straight cylindrical tube. However, working mechanisms of other shaking-up heat pipes with different geometries should be similar.

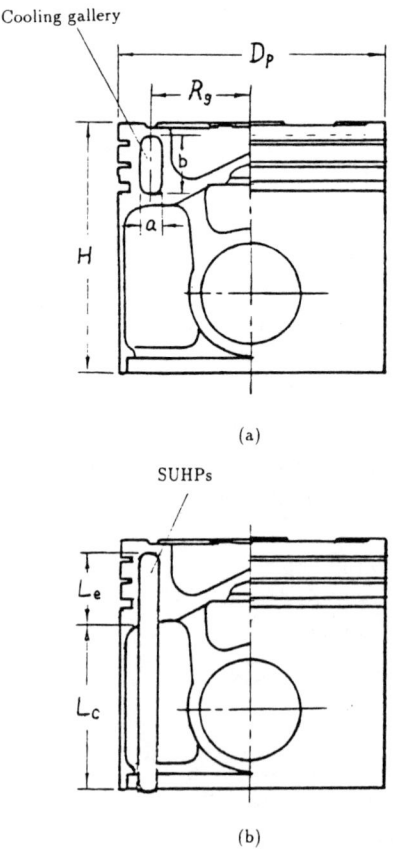

Figure 1: Schematics of gallery and SUHP cooled pistons.

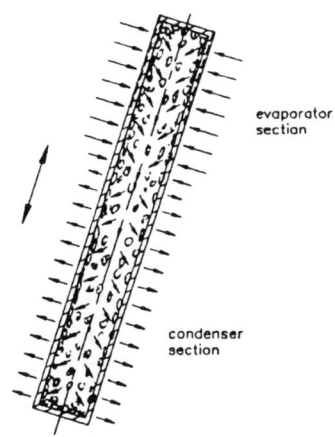

Figure 2: Schematic representation of the SUHP and interior working condition.

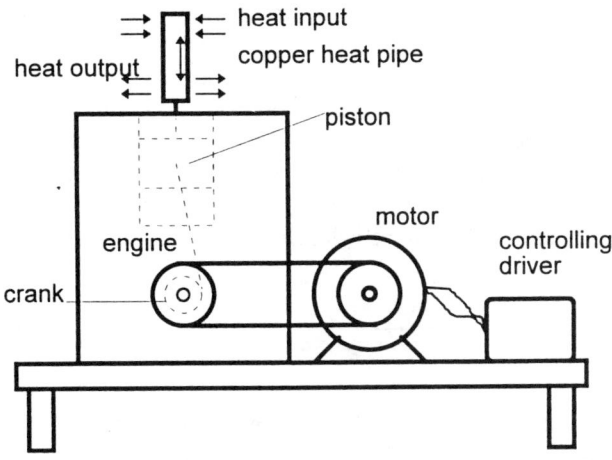

Figure 3: Schematic of the heat pipe test rig.

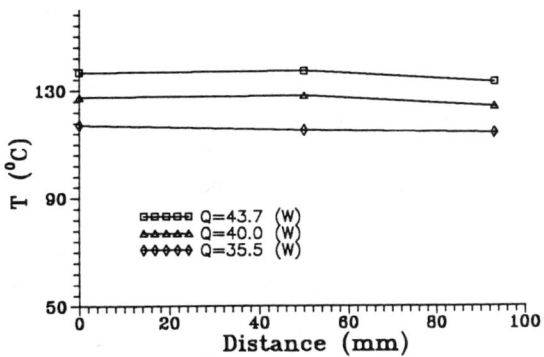

Figure 4: Temperature distribution along the heat pipe length at 8 Hz reciprocal motion.

THERMAL TESTING OF SHAKING-UP HEAT PIPES

As mentioned earlier, the major difference between the shaking-up heat pipe and other types of heat pipes is the liquid return mechanism. In order for a shaking-up heat pipe to work properly, the liquid inside the heat pipe should be able to reach the evaporator section constantly under the reciprocal motion of the heat pipe. For this purpose, an experimental observation using a transparent SUHP was performed by Wang and Cao (1994). This experimental observation revealed that when the engine cranking speed was around or above 420 rpm, a full scale liquid impingement on the upper evaporator wall was seen. Since the cranking speed at 420 rpm is well below the normal engine speed, the shaking-up heat pipe was proven to be feasible in an engine piston cooling application. A thermal test of a copper/water SUHP was also performed on a test rig (shown in Fig. 3). The test rig utilizes the true reciprocal motion of a single-cylinder internal combustion engine. The copper shaking-up heat pipe tested had an outer diameter of 10 mm, inner diameter of 8 mm, and a total length of 100 mm. The upper half of the heat pipe was wrapped with a flexible electric heater, which worked as an evaporator. Copper fins were attached at the lower half of the heat pipe, which worked as a condenser. The temperature distribution along the heat pipe was measured using 4 thermocouples, one of which was used to monitor the electric heater temperature. Figure 4 shows the temperature distributions along the heat pipe length for three different heat inputs at 8 Hz (480 rpm) reciprocal motion. As can be seen, the temperature is rather uniform along the heat pipe length. The different temperature levels at different heat loads is due solely to the cooling condition at the condenser section, which remained the same for different heat loads. The effective thermal conductances of the heat pipe for the three tested cases are about 35.0 to 37.0 times that of a solid copper bar having the same size as the SUHP, and 120.0 to 126.0 times that of the bar of Al-Si alloy used in the piston. In the above comparison, the thermal conductivities of copper and the Al-Si alloy were taken to be 400 (W/m-K) and 117 (W/m-k), respectively. Considering the small length of the heat pipe tested, the shaking-up heat pipe works remarkably well.

THERMAL ANALYSIS OF SUHP AND GALLERY COOLED PISTONS

In order to evaluate the effectiveness of shaking-up heat pipes in piston cooling, the performance of a cooling gallery is compared with that of the shaking-up heat pipes. The piston considered is a diesel engine piston with an outer diameter of D_b, and a height of H. The schematic configuration of the piston and the gallery are shown in Fig. 1a. For gallery cooling, a jet nozzle is used to inject oil into the gallery through an inlet port. The cooling oil entering the inlet port may flow out of the gallery at two or three outlet ports. Also, on the upward stroke, oil in the gallery dropped from the oil outlet ports, and only a very small amount of oil was retained in the gallery. To simplify the analysis, the oil flow is assumed to be a channel flow in the gallery along the piston circumference without multi-outlet port and oil dropping effects.

Since the oil gallery length is relatively short, and the thermal entry length for engine oil is large, the heat transfer of oil in the gallery should be in the thermal entry region. Using the concept of a hydraulic diameter, the mean Nusselt number is defined as:

$$Nu_m = h_m D_h / k \quad . \tag{1}$$

The values of the mean Nusselt number can be found from the dimensionless axial distance (Kays and Crawford, 1980), which is defined as:

$$L_g^+ = \frac{2(L_g / D_h)}{\text{Re}_h \text{Pr}} \tag{2}$$

where $\text{Re}_h = u_m D_h / \nu$ is the Reynolds number based on the hydraulic diameter of the gallery channel, u_m is the average oil velocity over the cross sectional area of the gallery channel, and L_g is the circumferential length of the gallery. The average oil velocity, u_m, can be calculated from:

$$u_m = \dot{m} / \rho A_c \tag{3}$$

where $\dot{m}$ is the oil flow rate in the gallery, and A_c is the cross-sectional area of the gallery. The oil flow rate in the gallery is related to the nozzle mass flow rate, $\dot{m}_n$, by:

$$\dot{m} = \eta \dot{m}_n \tag{4}$$

where η is the nozzle capture efficiency. The oil temperature at the gallery outlet, from Incropera and Dewitt (1990), is:

$$T_{out} = T_s - (T_s - T_{in}) \exp(-PL_g \bar{h} / \dot{m} c_p) \tag{5}$$

where T_s is the gallery wall surface temperature, T_{in} is the oil temperature at the gallery inlet port, P is the perimeter of the gallery cross section, L_g is the gallery length, $\dot{m}$ is the oil mass flow rate entering into the gallery, c_p is the specific heat of the oil, and $\bar{h}$ is the average heat transfer coefficient, which can be calculated from the mean Nusselt number:

$$\bar{h} = Nu_m k / D_h \tag{6}$$

The heat removed by the cooling oil in the gallery from the piston is:

$$Q_g = c_p \dot{m}(T_{out} - T_{in}) \tag{7}$$

For a typical gallery cooled piston with D_b = 130 mm, H = 140. mm, P = 71.3 mm, A_c = 0.264 × 10^{-3} m^2, L_g = 0.31 m, and D_h = 0.0148 m, if the gallery wall surface temperature is maintained at 200 °C, the oil inlet temperature, T_{in}, is 100 °C, the nozzle mass flow rate is 2.44 × 10^{-2} kg/s, and the nozzle capture coefficient is taken to be 1, the average heat transfer coefficient and the total heat removed from the piston, based upon Eqs. (1)-(7), are:

$$\bar{h} = 158.5 W / (m^2 - K) \tag{8}$$

and

$$Q_g = 324.4 W \tag{9}$$

For the same piston, the gallery is replaced with shaking-up heat pipes, as shown in Fig. 1b. In this case, the heat is transferred through the heat pipes from the upper piston region to the heat pipe condenser sections, and the jet nozzle impinges cooling oil directly onto the heat pipe condensers to take the heat away. The first step in calculating the heat transfer between the oil and the heat pipe is to compute the heat pipe thermal resistances through which the temperature drop, as well as the condenser temperature of the heat pipe, can be found. There are various heat pipe thermal resistances, which include those due to the container wall, vapor flow in the heat pipe core, and evaporation/condensation at the liquid-vapor interface. Due to the two-phase heat transfer nature of the shaking-up heat pipe and the relatively short length, only the thermal resistance due to the container wall is the most important one in this application (Dunn and Reay, 1982). Therefore, for a circular container, the thermal resistance for a shaking-up heat pipe can be calculated approximately as:

$$R_t = \frac{\ln(r_2 / r_1)}{2\pi L_c k} + \frac{\ln(r_2 / r_1)}{2\pi L_e k} \tag{10}$$

where r_2 and r_1 are the outer and inner radius of the heat pipe container, L_c and L_e are the condenser and evaporator lengths, and k is the container thermal conductivity. Since the thickness of the heat pipe container is very small, and the thermal conductivity of the container is large, the heat pipe thermal resistance should be very small. The heat pipe temperature in the condenser region, T_c, with a given temperature in the piston ring groove region, T_s, is:

$$T_c = T_s - (Q_{hp} / N) R_t \tag{11}$$

where Q_{hp} is the total heat transfer rate from upper piston region to the heat pipe condenser regions, and N is the number of the SUHPs used in the piston. For the shaking-up heat pipe cooled piston, the oil jet is used to impinge cooling oil to the lower (condenser) sections of the shaking-up heat pipes. For impingement heat transfer between the heat pipe condenser sections and the cooling oil, a correlation for the average Nusselt number based on the jet diameter due to Metzger et al. (1974) can be used:

$$\overline{Nu}_d = 2.74 \text{Re}_d^{0.348} \text{Pr}^{0.487} (2r/d)^{-0.774} (\mu_c / \mu)^{-0.37} \tag{12}$$

and the average heat transfer coefficient can be calculated by:

$$\bar{h} = \overline{Nu}_d k / d \quad (13)$$

where r is the radial distance from the stagnation point, *d* is the jet diameter, and μ_c and μ are the fluid dynamic viscosities at the wall surface and at the free stream, respectively. The Reynolds number is also based on the jet diameter, and is defined as:

$$\text{Re}_d = 4\dot{m} / \pi \mu d \quad (14)$$

where $\dot{m}$ is the oil mass flow rate that is impinged upon the heat pipe condenser section, which is related to the jet nozzle mass flow rate by:

$$\dot{m} = \eta_{hp} \dot{m}_n \quad (15)$$

where η_{hp} is the nozzle capture efficiency in this case. As mentioned in the description on the shaking-up heat pipe cooled piston, the condenser sections of the shaking-up heat pipes need to be placed in four circumferential segments in the lower piston section due to the existence of the wrist pin hole and the clearance required for oscillation of the piston connecting rod. For the oil impingement cooling here, each segment is considered to be a rectangular area to the jet nozzle, and is approximately equal to one-eighth that of the total skirt circumferential area:

$$A_{1/4} = \pi D_b L_c / 8 \quad (16)$$

In the formulation for the average heat transfer coefficient given by Metzger et al. (1974), *r* is the radial distance from the stagnation point. For a conservative consideration, 2r is taken to be the longer side of the rectangular area, which corresponds to a lower average heat transfer coefficient. Therefore, the total heat removed by the heat pipe is:

$$Q_{hp} = 4\bar{h} A_{1/4} (T_c - T_\infty) \quad (17)$$

where T_∞ is the oil free stream temperature that corresponds *to* the inlet temperature in the gallery cooling condition.

For the same piston dimension, the same piston temperature in the piston ring groove region, T_s, the same oil free stream temperature, and the same nozzle mass flow rate with a capture efficiency being equal to 1, as in the case for the gallery cooling, the average heat transfer coefficient, by using the above formulation, is:

$$\bar{h} = 1777.0 W / (m^2 - K) \quad (18)$$

and the ratio of the heat transfer rate using SUHPs to that using the cooling gallery is

$$Q_{hp} / Q_g = 8.07 \quad (19)$$

In the above calculation, the temperature drop from the upper evaporator section to the lower condenser section is assumed to be $10^0 C$, which is much higher than what would be obtained based on Eq. (10).

As can be seen, the heat removed by shaking-up heat pipes is more than eight times that removed by the cooling gallery based on the same ring groove temperature. This means that for the same piston ring groove temperature limit, the shaking-up heat pipe cooled piston can sustain much higher engine power. On the other hand, for the same amount of heat dissipation requirement, the shaking-up heat pipe cooled piston can maintain a much lower piston temperature. We should emphasize here that this conclusion based on the same heat dissipation rate is more practical than that based on a constant piston temperature, since the major heat transfer mode in the combustion chamber is radiation, which approximately depends on the fourth power of the combustion gas temperature. Also, since the piston temperature is always much lower than the engine combustion temperature that is on the order of a thousand degrees, the heat absorbed by the piston will remain virtually unchanged even if its temperature may decrease, say, from about $200\,^\circ C$ to about $150\,^\circ C$ due to the SUHP cooling effect. The analysis here, based on a constant piston temperature assumption, is merely to provide a vehicle for evaluating the cooling effectiveness of the shaking-up heat pipes.

We recognize that the above calculation is based on very simplified and approximate relations, and the results are qualitative in nature. However, the trends and magnitude of the comparison are valid. The major difference between the two cooling techniques is that the heat transfer coefficient of direct impingement cooling on the heat pipe condenser surface is an order of magnitude higher than that due to oil flow in the gallery, as is the case for many other heat transfer problems. Other practical factors may also favor the shaking-up heat pipe cooling method. In the above calculation, the nozzle capture efficiency for both cases is taken to be unity. However, the capture efficiencies for the cooling gallery should be much smaller that of the SUHP. In addition, the heat pipe cooling surface can be increased by attaching fins in the condenser region, which will contribute further to an increase in the cooling efficiency of the shaking-up heat pipes.

CONCLUSIONS

A piston cooling method employing shaking-up heat pipes has been described. The major difference between the shaking-up heat pipe and the existing heat pipes is the liquid return mechanism from the condenser section to the evaporator section, where the liquid splash and impingement on the inner surface play an important role. Experimental observation and thermal testing revealed that the SUHP is feasible in an engine piston cooling application. A simplified analysis on the piston cooling system indicated that for the same piston ring groove temperature limit, the shaking-up heat pipe cooled piston can sustain a much higher engine power compared to that with the gallery cooling system. On the other hand, for the same amount of heat dissipation requirement, the shaking-up heat pipe cooled piston can maintain a much lower piston temperature. The important feature of the SUHP cooled piston is to allow the engine to work at a higher temperature, which may contribute to both an increase in thermal efficiency and a reduction in environmental pollution.

ACKNOWLEDGMENT

The authors would like to thank Mrs. Helen Rooney at Florida International University for her help in preparing this manuscript.

REFERENCES

Cao, Y. and Wang, Q., 1994, ``A New Engine Piston,'' Patent Pending.

Chi, S. W., 1976, *Heat Pipe Theory and Practice*, Hemisphere Publishing, Washington D.C.

Dunn, P. D. and Reay, D. A., 1982, *Heat Pipes*, 3rd Edn., Pergamon Press, Oxford.

Incropera, F. P. and Dewitt, D. P., 1990, *Introduction to Heat Transfer*, John & Sons

Kays, W. M. and Crawford, M. E., 1980, *Convective Heat* and Mass *Transfer*, McGraw-Hill Book Company.

Metzger, D. E., Cummings, K. N., and Ruby, W. A., 1974, ``Effects of Prandtl Number on Heat Transfer Characteristics of Impinging Liquid Jets,'' *Proc. 5th Int. Heat Transfer Conf.*, Vol. 11, pp. 20-24.

Senda, J., Ogawa, T., Fujimoto, H,, Kubota, H., and Kimura, N., 1992, ``Characteristics of Combustion in an IDI Diesel Engine with a Swirl Chamber Made of Ceramics,'' SAE paper 920696.

Wang, Q., Cao, Y., and Souto, A., 1994, ``Development of a New Engine Piston Incorporating Heat Pipe Cooling Technology,'' to be presented at the 1995 Detroit SAE Congress.

950527

The Effect of the Addition of Hard Particles on the Wear of Liner and Ring Materials Running with High Sulfur Fuel

Jan Vatavuk and Valmir Demarchi
COFAP - Cia. Fabricadora de Pecas

ABSTRACT

The components of the piston/ring/liner system must have their wear resistance increased to meet the new engine requirements. The engine operating conditions can be even worse if corrosive wear in the engine is expected to occur.

This paper presents a study to improve the wear resistance of piston ring coatings and liner materials by the addition of chromium carbide and carbide forming alloying elements, respectively. The engine tests were run with high sulfur fuel (about 1.0 wt%) and lubricant with low total base number (TBN) with the objective of increasing the corrosive conditions.

The results show the improvement of the ring coatings wear resistance with the increase of the chromium carbide content. The cylinder liner materials also presented lower wear rates when they had hard particles, mainly due to the addition of niobium, vanadium and titanium.

INTRODUCTION

The heavy-duty diesel engines have been suffering higher loads due to the optimization of their performance by the engine manufacturers. The changes made in the engines increased the mechanical and thermal loads of their components, at the same time that their durability cannot be lowered due to the necessity of matching the exhaust emissions levels imposed by the legislation during the engine life.

The exhaust emissions levels of the diesel engines are largely influenced by the engine design characteristics, by its maintenance during its life, as by the fuel properties. Among the fuel properties, the sulfur content has been extensively studied. It is generally accepted that it influences the particulate emission [1] as well as the lubricant oil properties [2].

Although there is a tendency to use low sulfur fuels, there are some applications (shipboard engines) and some regions where the limits of the sulfur content are higher and only this type of fuel is available. In these cases, the wear of the components of the piston / ring / liner system is higher due to the corrosive effects of the combustion products [3]. Higher wear on the liner occurs mainly near the top dead center (TDC) of the first groove ring, because the temperature distribution and lubricating conditions on this area are favorable to the chemical attack of the combustion products. The corrosive working conditions are due to the condensation of the SO_2 and SO_3 formed during the combustion. When the presence of corrosive substances becomes higher than certain levels, the TBN of the lubricant cannot neutralize the acidity anymore.

The high wear of the cylinder liner, with the formation of polished areas (where the honing pattern is not present), leads to higher lubricant oil consumption, which also increases the particulate emission values [4][5]. For instance, the effect of the sulfur fuel content on the wear rate of the first groove rings and liners was determined in reference [6] for a specific engine for shipboard use, and the authors observed that increasing the fuel sulfur content from 0.4 to 1.5 wt% raised the top ring wear rate over nine times.

Besides the formation of polished areas, the localized wear on the region of the top ring motion reversion at the TDC is also important, as it can reach high values due to the operating conditions above discussed. In these conditions, the average bore wear rate can be 2 to 5 times higher than the normal rate, near the TDC of the first groove ring, as presented in reference [7].

For the same liner material, the polished area is directly related to the corrosive wear, but the TDC localized wear is mainly due to the combined effects of corrosive wear plus first groove ring coating material [8].

As the emission levels must be respected also during the engine life, the durability of the components of the piston / ring / liner system should be the highest possible, keeping the original engine design conditions, despite the higher loads applied to these components. In this way, the development of new materials and the improvement of their properties is gaining importance. The compatibility of the wear rates of the rings and the cylinder liners must be aimed, as the system has to be analyzed as a whole.

As the objective of this study is to analyze the wear resistance of new materials for liners and for ring coatings, the corrosive conditions in the dynamometric engine tests were increased by using high sulfur fuel (about 1.0 wt%) and low total base number (TBN) lubricant oil.

The study of the wear resistance improvement of the materials was carried out through the addition of chromium carbide (Cr_3C_2) hard particles to the plasma sprayed coating material of the top rings. Two different compositions were studied, containing 10% and 20% of chromium carbide added to a basic composition of molybdenum and a nickel rich microconstituent. On the other hand, the introduction of hard particles in the liner materials was done by the addition of carbide forming alloying elements. The studied materials, manufactured by centrifugal casting, were the following:

- normal production pearlitic gray cast iron with steadite;
- normal production with niobium, vanadium and titanium additions;
- normal production with niobium, vanadium, titanium and boron additions.

The wear results of these components will be presented and evaluated through dimensional variation and scanning electron microscopy analyses.

EXPERIMENTAL METHODS

WEAR MEASUREMENT - The dimensional variation due to the wear of the components was measured by metrological techniques. The variation of the ring gap was measured by putting the ring in a gage with the nominal bore diameter and the value was determined with the aid of a graduated conical lamina. The polished area was measured by cutting the liners through their diameter in the longitudinal direction, marking the polished areas and measuring them as a percentage of the total working area of the rings (from the TDC of the top ring to the BDC of the third ring). The localized wear at the TDC of the top ring was measured through a graph of the bore surface profile (in the axial direction, with high magnification), in relation to a reference line that links the worked and unworked regions, as shown in the example of figure 1.

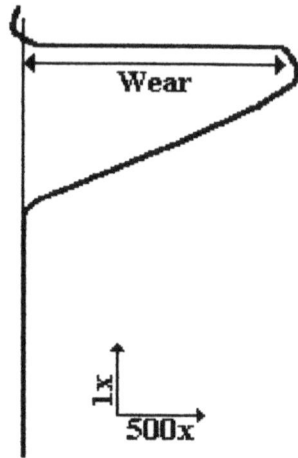

Figure 1: Scheme of the localized wear measurement.

ENGINE TESTS - The main characteristics of the engines used in the dynamometer tests are presented in table 1.

Table 1: Engine characteristics	
Engine type	Diesel cycle Four strokes 6 cylinders Turbocharged Direct injection 9.6 liters Truck application
Bore	120.65 mm
Maximum power	248 kW at 2050rpm
Maximum torque	1530 Nm at 1300rpm

The test procedure, shown in table 2, was determined based on a highway route where some wear problems were detected. Taking into account the road inclines, the load applied to the vehicle, the velocities, etc. a correspondent dynamometer test procedure was developed to simulate the road conditions in the test.

Table 2: Cyclic test procedure				
Item	Speed (rpm)	Torque (Nm)	Time (s)	Water out temp. (°C)
1	2050	1118	780	85
2	2270	255	60	58
3	1700	726	60	55
4	1300	1344	300	82
5	1000	49	300	48
6	1000	49	300	42

Before the cyclic procedure, the engines were run-in for 6 hours. The duration of the cyclic test was 400 hours and the tests were run in two engines to have a greater number of components to be analyzed.

The fuel used in the tests had a high sulfur content (about 1.0 wt%) and the lubricant oil was of API CD, SAE 20W20 grade, with a limited capacity of neutralizing the corrosive combustion products (TBN = 10 mg KOH/g, according to ASTM D2896).

It is important to mention that this type of engine has articulated pistons (steel crown and aluminum skirt). This type of piston allows better guidance of the crown, what minimizes the contact of the top land with the cylinder wall when the piston tilts. This characteristic assures us that the polished areas and the localized wear near the TDC of the first ring are due only to the ring contact, and not to the contact of the top land.

The cross sections of the rings are presented in Table 3.

Table 3: Ring pack cross sections	
Groove	Section
1st	
2nd	
3rd	

MATERIALS

<u>Cylinder liner materials</u> - Table 4 shows the chemical composition of the cylinder liner materials under analysis, where:

A - Normal production material;

B - Normal production plus Nb, V and Ti;

C - Normal production plus Nb, V, Ti and Boron.

The microstructure and hardness of the cast irons A and B are similar and are presented in figure 2.

Table 4: Chemical composition (wt%)											
	C	Si	Mn	P	S	Cr	Cu	Nb	V	Ti	B
A	3.20 3.50	1.90 2.40	0.60 0.80	0.60 0.80	0.05 max.	0.25 0.40	0.50 0.80	- -	- -	- -	- -
B	3.00 3.50	1.95 2.45	0.60 0.80	0.60 0.80	0.05 max.	0.30 0.50	0.50 0.80	0.25 0.35	0.20 0.30	0.05 max.	- -
C	3.00 3.50	1.95 2.45	0.60 0.80	0.60 0.80	0.05 max.	0.30 0.50	0.50 0.80	0.25 0.35	0.20 0.30	0.05 max.	0.06 max.

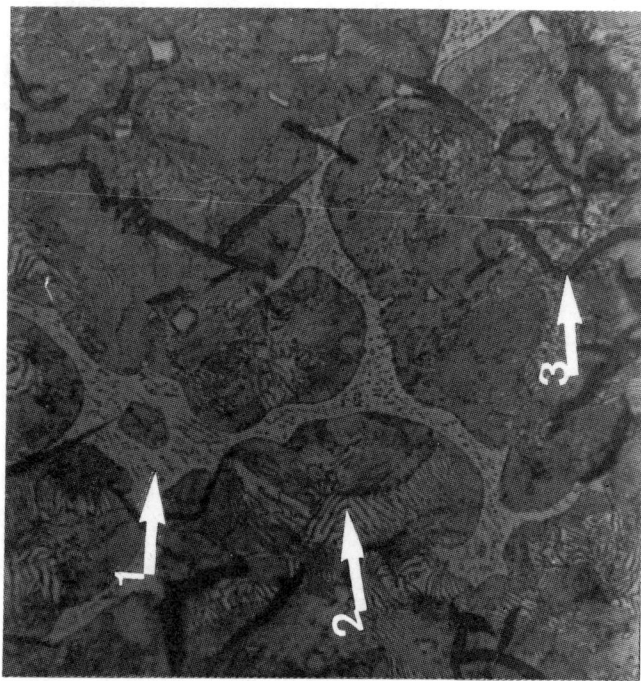

Figure 2: Microstructure representative of materials A and B: steadite (arrow 1), pearlite (arrow 2), graphite (arrow 3); 260 $HB_{187.5}$ with 2.5 mm diameter sphere; 700x magnification.

The MC niobium, vanadium and titanium rich hard particles of materials B and C are similar and can be seen in figure 3. The titanium addition was done just to modify the morphology of the MC type carbides (niobium and vanadium rich) from the Chinese script one to a compact morphology.

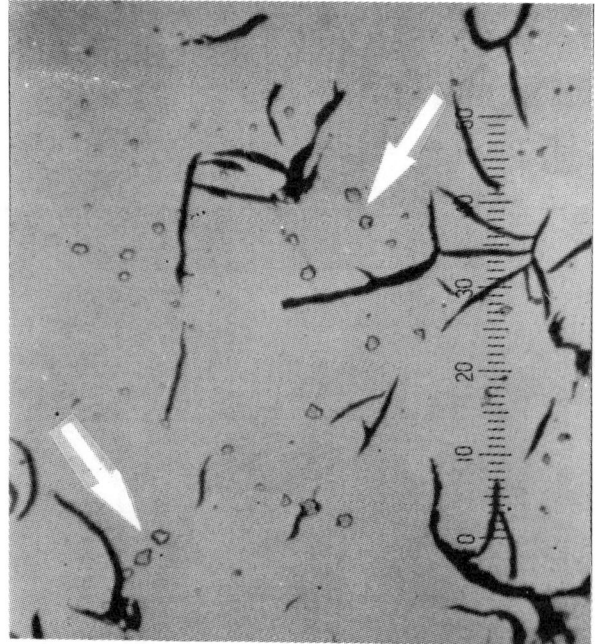

Figure 3: MC particles aspect in materials B and C; 400x magnification.

Although the boron addition does not change the size, shape and distribution of the MC particles, it changes the shape and increases the volume fraction of hard particles of the steadite microconstituent. Figure 4 shows the microstructure and hardness of material C.

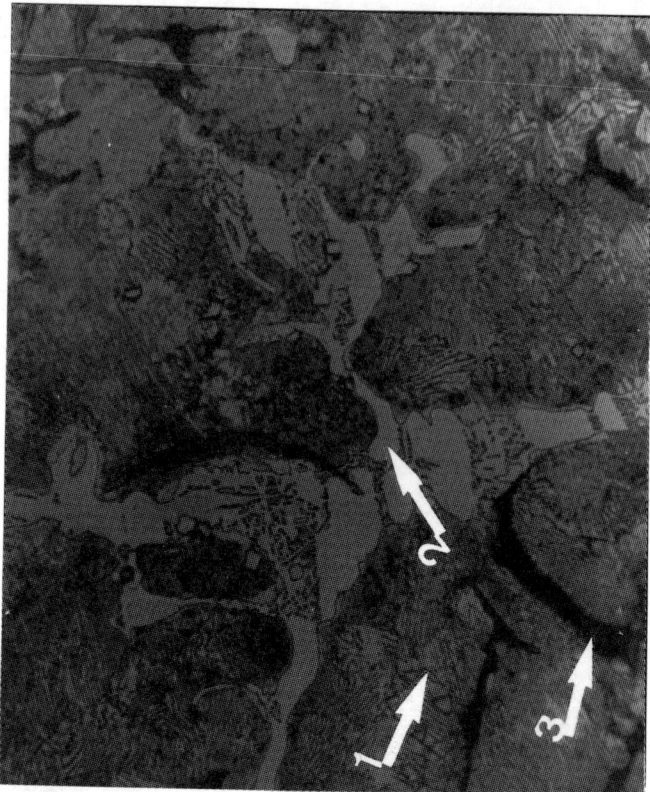

Figure 4: Microstructure of material C: pearlite (arrow 1), steadite (arrow 2) and graphite (arrow 3). 270 $HB_{187.5}$ with 2.5 mm diameter sphere; 700x magnification.

The higher volume fraction of hard particles in material C is due to a higher Fe_3C volume fraction in the steadite microconstituent.

The mechanical features of the microconstituents of materials A, B and C, measured by microhardness indentation, are presented in table 5.

Table 5: Microconstituents Vickers hardness with 25 g load		
Microconst.	Material	HV
pearlite	A, B and C	360
steadite	A and B	900
steadite	C	1000
(Standard deviation smaller than 20% of the average value)		

The pearlite hardness was the same for the three materials, even with the boron presence. The higher hardness of the steadite in material C is related to its more compact structure if compared with the steadite of materials A and B.

The MC hardness was not measured because of its small size, but it can be inferred from the literature as around 2000 HV.

<u>First groove piston ring coating materials</u> - These coatings were obtained by the thermal plasma spraying process. After applying the coatings, they are finished with a fine grind on their working surface, respecting the dimensional requirements of the finished product.

Table 6 shows the chemical composition and percentages of the powder particles that result in the two Cr_3C_2 grades coatings under study.

Table 6: Chemical composition (wt%) of coatings C1 and C2	
C1	70% Mo 20% Ni rich microconstituent 10% Cr_3C_2
C2	64% Mo 16% Ni rich microconstituent 20% Cr_3C_2
Ni rich microconstituent	15% Cr; 3.5% B; 4.5% Si; 4.5% Fe; 0.9% C; Ni balance

The powder particles to be sprayed have sizes between 20 and 77 µm. Figures 5 and 6 present the coatings C1 and C2 microstructures seen by a back scattering image in the scanning electron microscope.

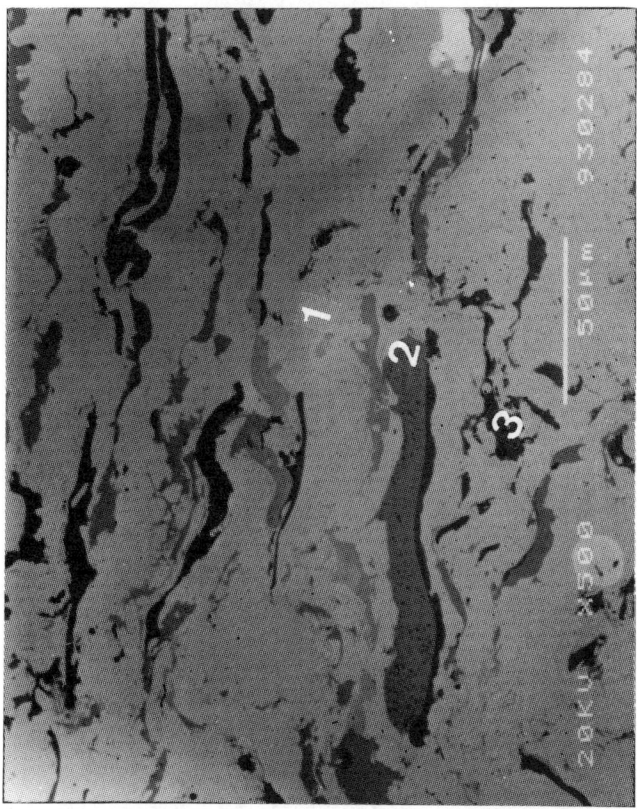

Figure 5: Coating C1: the white area is the Mo {1}, the darker area is the Ni rich microconstituent {2} and the darkest one represents the Cr_3C_2 {3}.

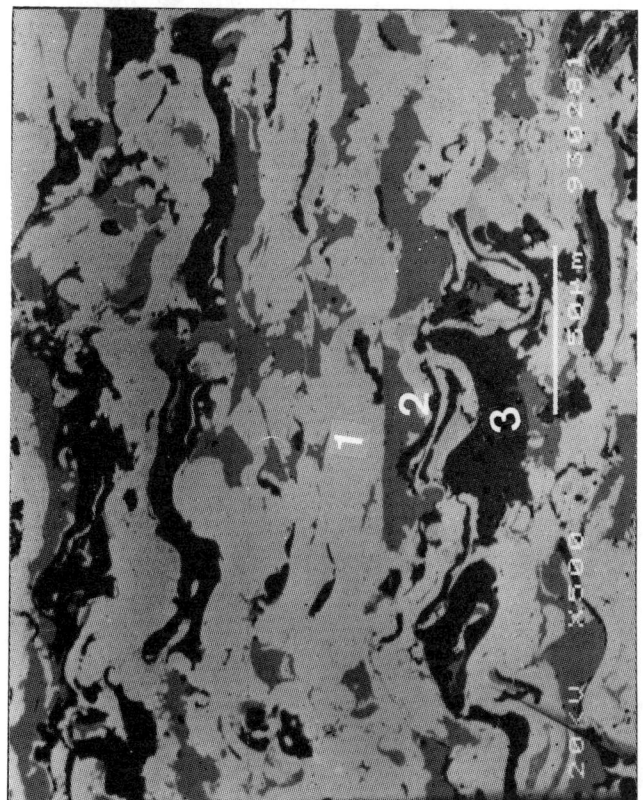

Figure 6: Coating C2: the same as figure 5.

The main difference between C1 and C2 is the higher chromium carbide content of the latter. The mechanical characteristics of the coatings measured by microhardness indentation can be seen in table 7.

Table 7: Coating Vickers hardness with 25g load.

Mo	700
Ni rich microconstituent	900
Cr_3C_2	1700
(Standard deviation smaller than 20% of the average value)	

The second and third groove rings were the normal production type, with an electrolytic chromium plating with 1000 HV.

ENGINE ASSEMBLY - In order to minimize the influence of the temperature profile through the engine, due to the cooling water inlet and outlet locations, the different materials of the liners and top ring coatings were arranged as shown in table 8.

Table 8: Materials distribution in the engine

Test	Material	Cylinder					
		1	2	3	4	5	6
1	Liner	B	C	A	A	C	B
	Top ring	C2	C1	C2	C1	C2	C1
2	Liner	B	C	A	A	C	B
	Top ring	C1	C2	C1	C2	C1	C2

RESULTS AND DISCUSSION

ENGINE TESTS RESULTS - The oil consumption and blow-by averages, measured during the 400 hour tests, are presented in table 9. These values can be considered normal for this type of engine and test procedure.

Table 9: Oil consumption and blow-by results

Test	Oil consumption (g/h)	Blow-by (L/s) at the 2050 rpm step
1	77.3	0.875
2	107.9	0.692

The TBN of the lubricant oil during the tests varied from 10.0 mg KOH/g for the new oil to about 4.0 mg KOH/g after 200 hours when the oil was changed. In these tests, the metallic particles content in the oil was measured and the values can be considered normal, showing that the wear rates were coherent considering the good engine functioning.

WEAR RESULTS

<u>Cylinder liners</u> - The honing characteristics of the cylinder liners were previously selected to make sure that all the liners, even being made of different materials, had similar surface finishing, since these characteristics are very important to determine the wear rates of the rings and liners.

The cylinder liner wear values are presented in table 10. It shows the average localized wear at the TDC of the top ring (four measurements each liner), its maximum values (for both tests) and the average polished area for each combination of liner material and top ring coating material.

As we can see in table 10, the localized wear at TDC of the first groove ring, as well as the polished area, have the minimum values for material B running against the two types of coatings. The cylinder liners made of material C have intermediate behavior if compared with materials A and B. Figures 7 to 9 show the maximum localized wear at TDC for the three materials.

Liner material	C1 top ring coating			C2 top ring coating		
	Average TDC wear (µm)	Max. TDC wear (µm)	Average polished area (%)	Average TDC wear (µm)	Max. TDC wear (µm)	Average polished area (%)
A	19.1	42.0 / 30.0	11.0	28.7	60.0 / 39.0	10.7
B	8.3	19.5 / 9.0	4.1	14.9	33.0 / 18.0	5.3
C	11.5	28.5 / 15.0	7.9	17.8	45.0 / 18.0	7.0

Table 10: Cylinder liner wear

Figure 7: Secondary electron image of the maximum localized wear for material A (test 1) running against coating C2.

Figure 8: The same as figure 7, for material B (test 1) running against coating C2.

Figure 9: The same as figure 7 for material C (test 1) running against coating C2.

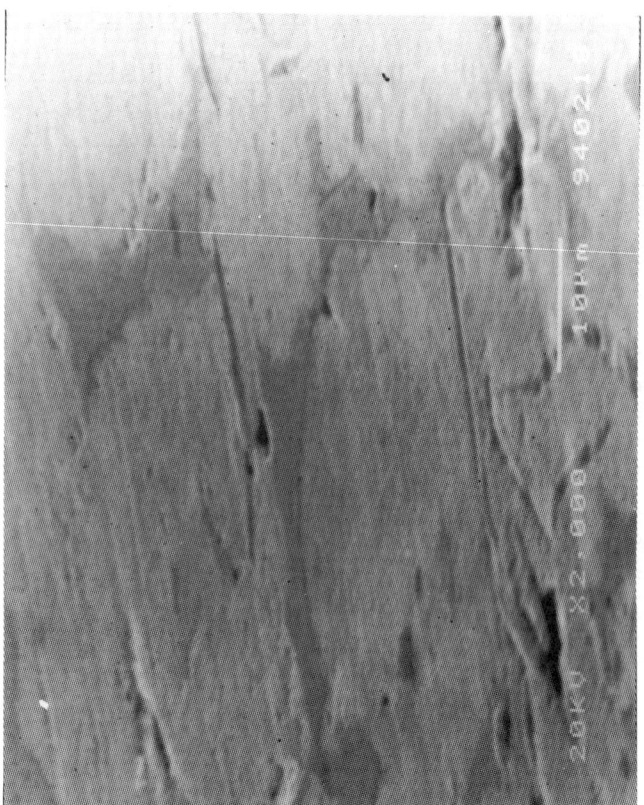

Figure 10: Secondary electron image of the cylinder liner of material A, 25 mm from the top ring TDC. Steadite is protruding.

The best results were obtained with the smallest microstructure change, from material A to B, which have as the only difference a small volume fraction of MC type carbides, with a small compact morphology, obtained by the addition of Nb, V and Ti elements. The wear pattern seen on material A in figure 7 shows a very steep aspect at the borderline of the worked and unworked cylinder liner regions. The same pattern should be expected to occur on material B for a long term test.

Although material C has the highest intermetallic volume fraction, which is a boron effect on the steadite (figures 10 and 11), and the same addition of MC type carbide forming elements as material B, its behavior is not as good as the latter.

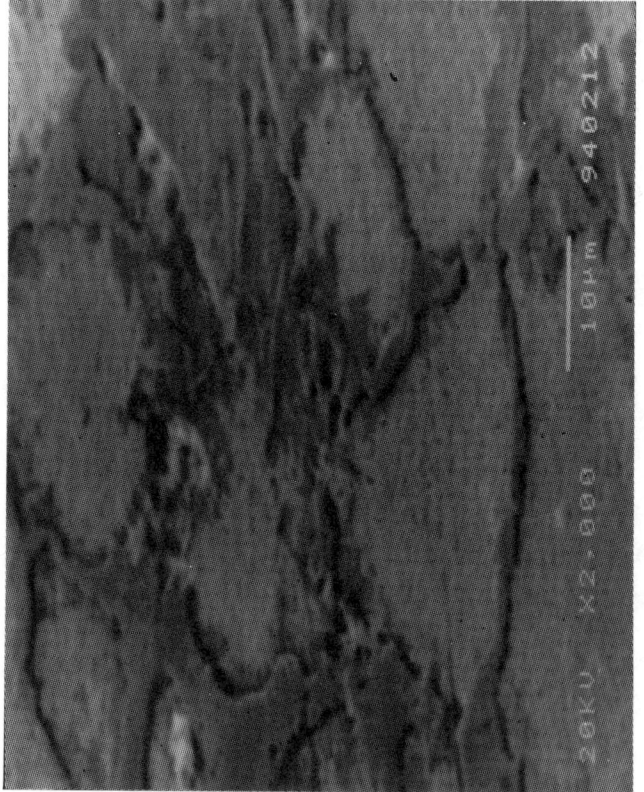

Figure 11: The same as figure 10 for material C.

Table 10 shows also that the TDC wear of the cylinder liners was systematically higher when the first groove piston ring coating was changed from material C1 to C2. However, the polished area did not show the same tendency, presenting similar values running against both coating materials, showing no influence of the Cr_3C_2 volume fraction increase.

Even comparing material C2 with a coating without any Cr_3C_2, which was the base for the development of C1 and C2, i.e., the same molybdenum and nickel rich microconstituent ratio, the polished area did not show any difference in an engine test under the same conditions [8]. It means that the Cr_3C_2 addition has no effect on the polished area, but is effective in enhancing the first groove piston ring coating wear resistance, as it will be discussed in the next item.

The polished area is a function of the corrosive wear particles action, such as detached steadite near the first groove ring TDC [7][8], whose action depends on the lubricant film thickness behavior below the movement reversion of the top ring. However, the localized wear at the TDC of the top ring depends on the coating characteristics (even corrosive wear resistance properties) because of the higher probability of metallic contact between the ring coating and the liner in this region or a little below it due to the squeeze effect.

The lubricant / fuel properties are important to the mechanisms above mentioned, that is, a flat viscosity/temperature curve for the lubricant oil can increase the film thickness at high temperatures near the combustion chamber, and also reduce the probability of metallic contact in the vicinity of the movement reversion of the top ring. Likewise, a high TBN associated with a low sulfur fuel, can reduce the corrosive wear conditions, with lesser detachment of particles.

Piston rings - The wear of the second and third groove piston rings after both 400 hour tests can be seen in table 11.

Table 11: Gap increase of second and third groove rings (mm)		
Liner material	2nd ring	3rd ring
A	0.08	0.12
B	0.07	0.13
C	0.07	0.11
(Each value corresponds to an average of four rings)		

Taking into account that all the 2nd and 3rd groove rings have the same type of coating (chromium plating), the wear results shown above indicate similar behavior among the different cylinder liner materials against the rings.

Table 12 presents the first groove ring coating wear measured by the gap increase of the rings for both tests.

Table 12: Gap increase of the first groove ring (mm)		
Liner material	Ring coating	
	C1	C2
A	0.175	0.175
B	0.180	0.120
C	0.170	0.140
Average	0.175	0.145
(Each value corresponds to an average of two rings)		

Analyzing the results shown in table 12, we can note that coating C2 has higher wear resistance in relation to C1. This means that a higher volume fraction of Cr_3C_2 particles improves the wear resistance of these first groove piston rings. No effect of the cylinder liner material on the top ring wear could be noted, just as in the case of the second and third groove rings.

Combined wear: Based on the results shown in the previous tables (10 to 12), figure 12 presents the wear results of the pair first groove ring coating / cylinder liner, to achieve the best tribological pair configuration.

As shown in figure 12, the best combination of materials in this study is coating C2 with cylinder liner material B.

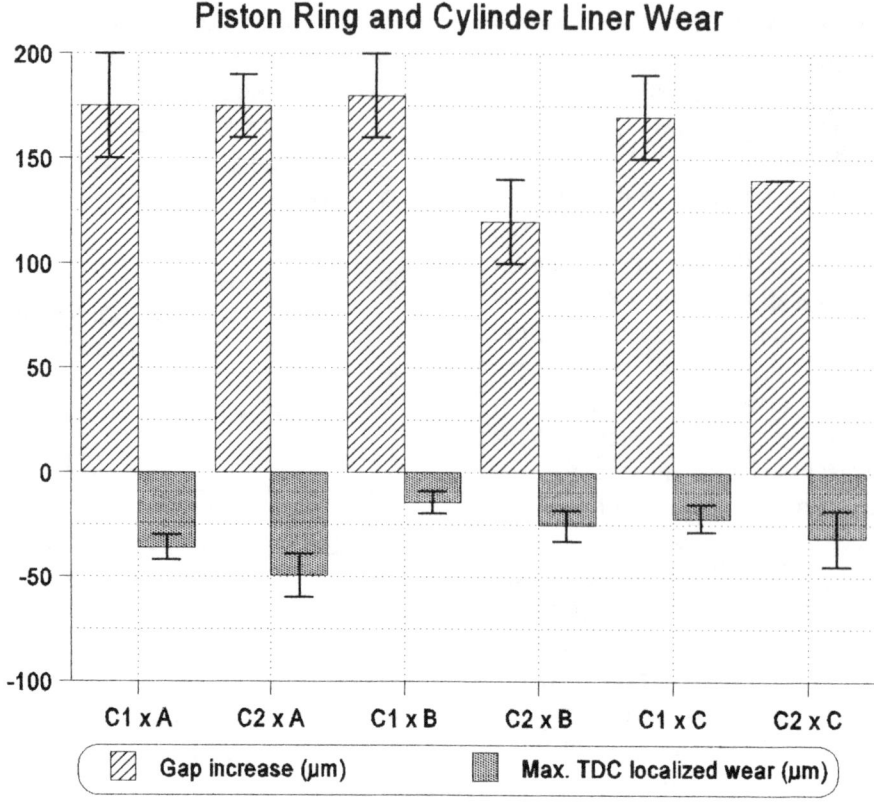

*Figure 12: Piston ring and cylinder liner wear
(the indications at the bar ends represent the values of both tests).*

CONCLUSION

◊ The polished area as well as the localized wear at the TDC of the first groove ring were smaller for the material containing niobium, vanadium and titanium additions (material B). The material with these additions plus boron (material C) had intermediate behavior, between the normal production (material A) and the previously mentioned material B, in spite of its higher volume fraction of steadite in relation to materials A and B.

◊ Increasing the Cr_3C_2 content of the coatings raises the localized wear at the top ring TDC for the three cylinder liner materials. However, the polished area had no significant variation.

◊ The increase of the Cr_3C_2 content reduces the wear rate of the piston ring coating materials with molybdenum and nickel rich microconstituent basic composition.

◊ The chromium plated second and third groove rings presented no significant wear variation running against the three different cylinder liner materials. The same fact was observed for the first groove ring.

◊ The best pair of materials, showing the lowest combined wear, was first groove ring coating C2 and cylinder liner material B, containing Nb, V and Ti additions.

REFERENCES

[1] Mullins, P. "Are New Diesel Fuels Needed for Cleaner Air?". High Speed Diesels & Drives - March, 1994. p. 36-37.

[2] Ripple, D.E. and Guzauskas, J.F. "Fuel Sulfur Effects on Diesel Engine Lubrication". SAE paper nr. 902175.

[3] Schramm, J.; Henningsen, S. and Sorenson, S.C. "Modelling of Corrosion of Cylinder Liner in Diesel Engines Caused by Sulphur in the Diesel Fuel". SAE paper nr. 940818.

[4] Zelenka, P.; Kriegler, W.; Herzog, P.L. and Cartellieri, W.P. "Ways Toward the Clean Heavy-Duty Diesel". SAE paper nr. 900602.

[5] Ishizuki, Y.; Sato, F. and Takase, K. "Effect of Cylinder Liner Wear on Oil Consumption in Heavy-Duty Diesel Engines". SAE paper nr. 810931.

[6] Balnaves, M. et al. "Fuel Property Effects on Ring and Liner Wear Rates in a DDC 6V-53T Using SLA Techniques". SAE paper nr. 912326.

[7] Aeberli, K. and Lustgarten, G.A. "Verbessertes Kolenlaufverhalten bei Langsamlaufendem"; Sulzer Dieselmotoren, MTZ, - May, 1989. p. 197-204.

[8] Vatavuk, J. "Mecanismos de Desgaste em Anéis de Pistão e Cilindros de Motores de Combustão Interna" ("Wear Mechanisms of Piston Rings and Cylinder Liners of Internal Combustion Engines"). Ph.D. Thesis - Escola Politécnica da Universidade de São Paulo, 1994.